Dorin Catana

Risk Factors Combating of the Work Environment

AF138165

Dorin Catana

Risk Factors Combating of the Work Environment

LAP LAMBERT Academic Publishing

Impressum / Imprint

Bibliografische Information der Deutschen Nationalbibliothek: Die Deutsche Nationalbibliothek verzeichnet diese Publikation in der Deutschen Nationalbibliografie; detaillierte bibliografische Daten sind im Internet über http://dnb.d-nb.de abrufbar.
Alle in diesem Buch genannten Marken und Produktnamen unterliegen warenzeichen-, marken- oder patentrechtlichem Schutz bzw. sind Warenzeichen oder eingetragene Warenzeichen der jeweiligen Inhaber. Die Wiedergabe von Marken, Produktnamen, Gebrauchsnamen, Handelsnamen, Warenbezeichnungen u.s.w. in diesem Werk berechtigt auch ohne besondere Kennzeichnung nicht zu der Annahme, dass solche Namen im Sinne der Warenzeichen- und Markenschutzgesetzgebung als frei zu betrachten wären und daher von jedermann benutzt werden dürften.

Bibliographic information published by the Deutsche Nationalbibliothek: The Deutsche Nationalbibliothek lists this publication in the Deutsche Nationalbibliografie; detailed bibliographic data are available in the Internet at http://dnb.d-nb.de.
Any brand names and product names mentioned in this book are subject to trademark, brand or patent protection and are trademarks or registered trademarks of their respective holders. The use of brand names, product names, common names, trade names, product descriptions etc. even without a particular marking in this works is in no way to be construed to mean that such names may be regarded as unrestricted in respect of trademark and brand protection legislation and could thus be used by anyone.

Coverbild / Cover image: www.ingimage.com

Verlag / Publisher:
LAP LAMBERT Academic Publishing
ist ein Imprint der / is a trademark of
OmniScriptum GmbH & Co. KG
Heinrich-Böcking-Str. 6-8, 66121 Saarbrücken, Deutschland / Germany
Email: info@lap-publishing.com

Herstellung: siehe letzte Seite /
Printed at: see last page
ISBN: 978-3-659-26421-4

FOREWORD

Occupational Safety and Health is a cross-disciplinary area concerned with protecting the safety, health and welfare of people engaged in work or employment. The goal of all occupational safety and health program is to foster a safe work environment. As a secondary effect, it may also protect co-workers, family members, employers, customers, suppliers, nearby communities, and other members of the public who are impacted by the workplace environment.

An effective training program can reduce the number of injuries and deaths, property damage, legal liability, illnesses, workers' compensation claims, and missed time from work. Safety training classes help establish a safety culture in which employees themselves help promote proper safety procedures while on the job. It is important that new employees be properly trained and to embrace the importance of workplace safety.

In the European Union, member states have enforcing authorities to ensure that the basic legal requirements relating to occupational health and safety are met. In many EU countries, there is strong cooperation between employer and worker organization (e.g. Unions) to ensure good OSH performance as it is recognized this has benefits for both the worker (through maintenance of health) and the enterprise (through improved productivity and quality). In 1996 the **European Agency for Safety and Health at Work** was founded.

Member states of the European Union have all transposed into their national legislation a series of directives that establish minimum standards on

1

occupational health and safety. These directives follow a similar structure requiring the employer to assess the workplace risks and put in place preventive measures based on a hierarchy of control. This hierarchy starts with elimination of the hazard and ends with individual protective means (personal protective equipment). The terminology used in OSH varies between states, but generally speaking:

- a hazard is something that can cause harm if not controlled;
- the outcome is the harm that results from an uncontrolled hazard;
- a risk is a combination of the probability that a particular outcome will occur and the severity of the harm involved.

Hazard analysis or hazard assessment is a process in which individual hazards of the workplace are identified, assessed and controlled/eliminated as close to source (location of the hazard). Permanently must understand that risk assessment requires as the hazard to be established to a reduced level but practically acceptable.

The paper presents synthetic, in an accessibly technical language, the knowledge necessary to combating of the risk factors that belong to the work environment.

In time is possible as the society mentality represented by employers and employees to change, so the attention will be offered to field to become higher, concomitant with understanding the benefits generating by the respecting and application of the field specific norms.

There should be hope that society attitude will change gradually from the stage in which the occupational safety and health field settlement respecting is essential to that in which the preventive aspects of this activity will be a priority.

Author

TABLE OF CONTENTS

*For all those that made possible
the writing and publishing this book*

RISK FACTORS COMBATING
OF THE WORK ENVIRONMENT

1   Accidents at work and occupational illnesses

The concept of **accident** generally designates an unexpected event that occurs suddenly is unpredictable and interrupt the normal progress of activity. The concept of work accident must associate to a work process and compulsory involve human factors. The work accidents consist in the biological component injury of the human factor, unexpectedly, suddenly and violently, in the course of a labour process. It should be noted that the definition of work accident, just surprise appearance of the biological component injury, not including the mental side of human personality. The recent researches plead for extending the concept of work accidents to the aspects regarding the psychological component damage.

The term "work accident" is used always in reference to the human factor who suffered some degradation in his work ability or death, in terms of his professional activity and because of it. In order to reduce the number of accidents, the International Labour Office (BIT) initiates continuous research and proclaim measures to ensure the safety of workers. For all BIT Member States, these laws are integrated into national legislation.

The improving of occupational safety and health is a central issue for the Member States of the European Union that under Directive 89/391/EEC [12] developed a continuous improvement of legislation, designed to ensure prevention of accidents and occupational being taken ill as well and better conditions of work, for all categories of workers.

Among the work accident characteristics can be mentioned: the unpredictability, the causes and effects, actions or inactions of the performer at a time. Specific is the disproportion between cause and effect, most often small causes produce great effects. Most of the accidents occur due to negligence, the mistakes of perception, information incorrect processing. As defined by the World Health Organization, occupational illnesses are affections whose specific etiologic agents are present in the workplace, associated with some (certain) industrial operations or a profession.

In view of the above, as a condition of the body injury (disease) to be qualified as an occupational disease must be satisfied the following conditions:

- to appear from the practising of a trade or profession;
- be caused by harmful factors, physical, chemical, biological, characteristics of the workplace, or of overload.

In the worker exposure case to harmful factors, the pathological process is slow and affects either the overall condition or certain systems, organs. Often, the damage produced is reversible. Through interrupting the action of harmful factors on patient and, adequate treatment applying, the consequences of disease are attenuated or disappear completely. For a noxious (harmful factors, risk factor that cause being taken ill), of the working environment to be recognized as an occupational illnesses etiologic factor, there must be evidence of a quantitative relationship between the contaminant dose absorbed by the body and the effect on it. The dose-effect relation was determined for a large number of harmful factors, imposing maximum limits.

The assessment modes of the work safety, in a work system can be most clearly differentiated by the moment when the assessment is made, in relation to the events that define the presence or absence of safety – accidents and occupational being taken ill (disease). From this point of view, there are two **principles of safety assessment:**

- **post-accident/occupational being taken ill;**
- **ante-accident/occupational being taken ill.**

Depending on the principle adopted, the assessment criteria are differentiated:

- **number and severity of the actually events produced;**
- **probability and potential seriousness of the events that might occur in the system.**

Post-accident assessment allows the estimate security degree of the work in a production system, based only on occupational accidents and illnesses product into the system, in a certain period of time. Figure 1 presents the evolution of accidents at work in most industrialized countries of the European Union from 1995 to 2007 [7, 15].

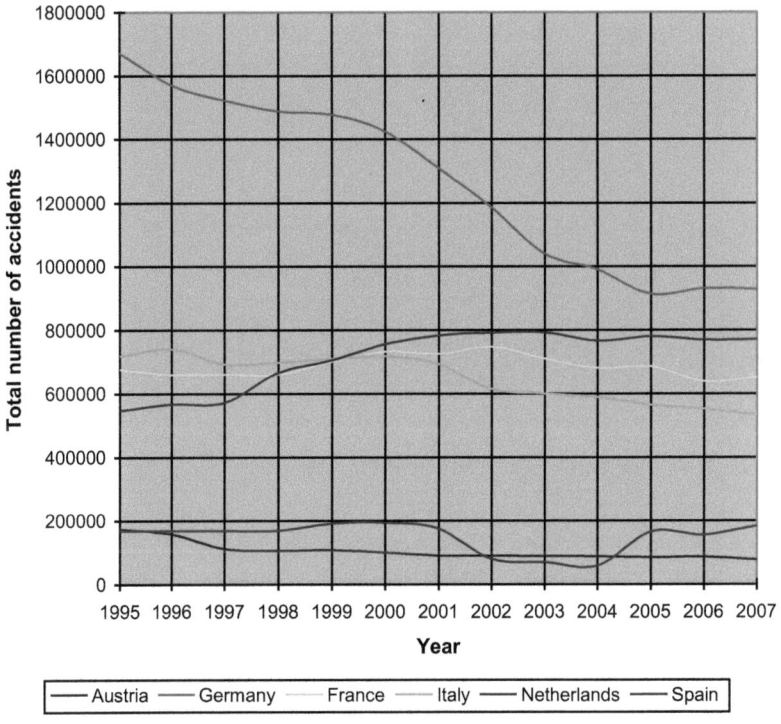

Fig. 1 *Number of accidents (more than 3 days sick leave)*
in few countries of Europe

7

In most countries the number of work accidents diminished, Spain is the only country where, to the end of the period showed a higher number of accidents (771,014 work accidents in 2007 compared to 547,003 in 1995).

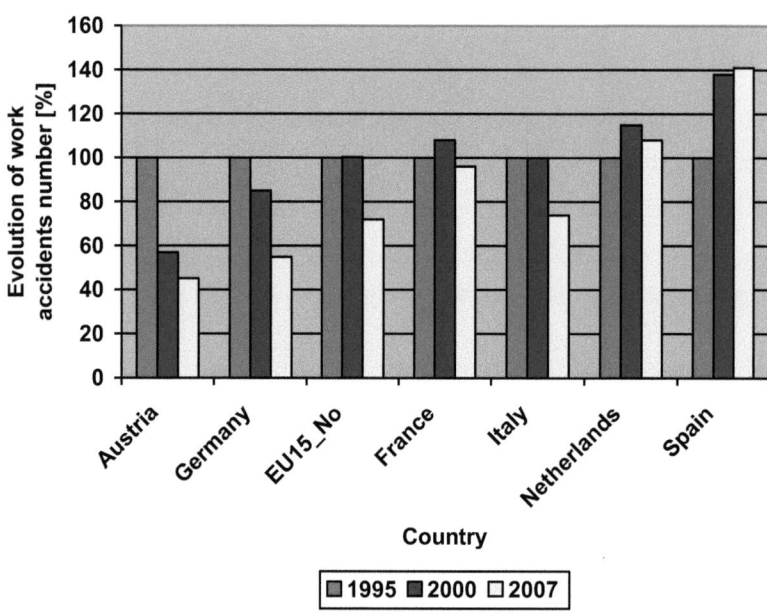

Fig. 2 *Evolution in percentages of the accidents number*
1995 - 2007 (1995 base year)

Figure 2 presents the percentage evolution of the accidents number in the period 1995-2007. Analyzing the figure, it appears that in the European Union (EU15_No – 15 countries plus Norway) accidents number decreased by 28% during the mentioned period. Significant reductions by far under the European average were recorded in Austria and Germany, where the accidents number in 2007 is approximately 45% and 55% compared to 1995 (which means a reduction of 55% and 45% of accidents). In Spain the accidents number in the analyzed period increased by 41% [1, 2]. For comparative evaluation of the work security degree in the analyzed system are used two types of statistical indicators: absolute and relative.

Absolute indicators express in absolute measure the number of accidents/illnesses product without to be realized their reporting to other sizes. They indicate in a given period, the dynamics level and, structure of occupational accidents and/or illnesses in the analyzed system. In this category are:

- total number of accidents or illnesses products (fatal, with invalidity, with work temporary incapacity);
  - total number of collective work accidents;
  - total number of days, of work temporary incapacity;
  - assistance cost for temporary incapacity.

Absolute indicators permit to characterize the overall situation of work safety, without the possibility to make the deeper evaluations and comparisons.

Relative indicators expressing the number of accidents/illnesses caused at the system level, reported to other sizes, allowing the effectuation a more conclusive comparisons.

Frequency index and the **gravity (severity) index** are the most used in the analysis of workplace accidents.

Frequency index (I_{fa}) indicates the number of accidents per 1,000 employees and is calculated by the relation 1:

$$I_{fa} = \frac{N_a}{N_s} \cdot 1,000 \quad , \tag{1}$$

where:

N_a – total number of wounded persons;

N_s – average number of personnel during the period of the reporting of the accidents.

Using relation 1, frequency index can be calculated for all accidents: fatal, with invalidity or with work temporary incapacity. The figure 3 shows the number of serious accidents (frequency index) per 100,000 employees in most industrialized countries of the European Union.

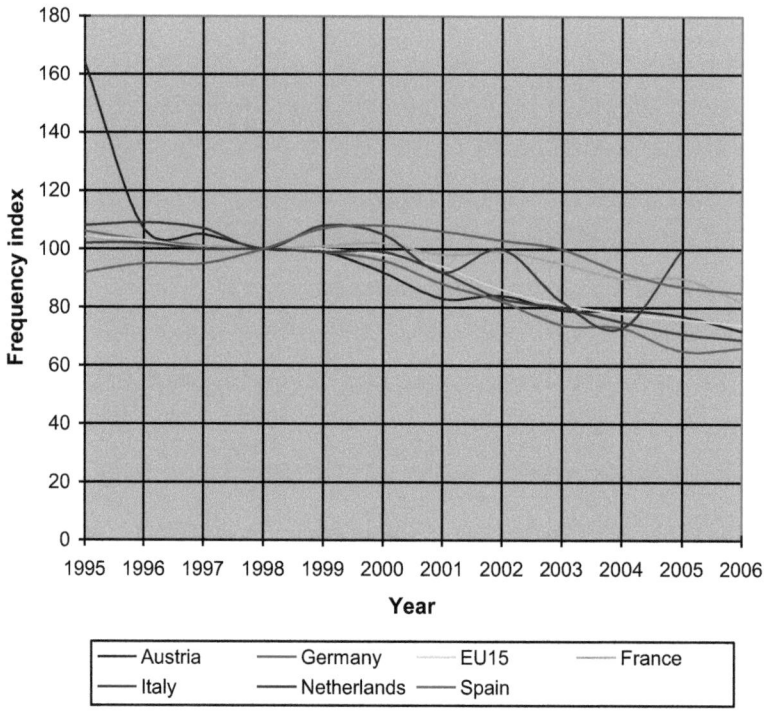

Fig. 3 *Number of fatal (serious) work accidents incumbent*
100,000 employees

The graph of figure 3 encourages the conclusion obtained by analyzing the figure 1, through it was established that the European Union countries level, the number of work accidents decreased from 1995 -2007. So, this is visible by frequency index of work accidents.

Gravity (severity) index (I_{ga}) indicates the number of man-days work incapacity owed to work accidents per 1,000 employees and is calculated by the relation 2:

$$I_{ga} = \frac{N_i}{N_s} \cdot 1,000 \quad , \tag{2}$$

where:

N_i – calendar days number of work incapacity;

N_s – average number of personnel during the period of the reporting of the accidents.

When the reference time is less than one year, for frequency and severity indices are applied a correction that allows their comparison with annual levels. In conclusion, the fundamental concepts used in post-accident assessment of the work safety into a system are the accidents and illnesses already produced; the assessment criteria, principal refer to the frequency and severity rate, and the method used is **the statistical analysis of the occupational accidents and illnesses.**

Main limits of post-accident assessment can be summarized as:

- not take into account potential situations of accident and illness;

- can not be applied to the design level of production systems;

- not possible to define the risk/security level, of work in the new objectives, set into operation recently, where are not yet accidents or occupational illnesses, but may present a great potential, of danger;

- main scope is to establishment of the occurred events and no to prevention.

Ante-accident assessment permits the identification and establishment of the risk factors of the injury or occupational being taken ill, in the work systems. Once, these identified, can be taken measures to control or reduce them. The preventive measures of the risks assessment that can produce work accidents and occupational illnesses, permits the considerable reduction of these, in the European countries, in 1995-2007 period (see fig. 2), corroborated with the employer and employee understanding the importance, of this mision.

Further, are presented the combating methods of the most important risk factors of the work environment, encountered in the work systems.

2   Industrial ventilation

The air characteristics (temperature, humidity, air current, pressure) of the work environment represent physical risk factors for operators. The air of the workspaces is continuous vitiated because of the gases emissions, vapours, powders, dust and the heat generated by the work processes. In the workplaces where such conditions appear, the health of workers suffer, productivity decreases and the raw materials, finished products and other materials can degrade.

The ventilation is the main measure of labour protection to eliminate noxiousness of the workplace air. It is the process of replacing the vitiated or contaminated air from chambers, with fresh air from outside or with purified air [8]. Also with the ventilation support can be realized the control of microclimate factors: temperature, humidity, and air velocity. It is possible the development of ventilation systems for the completely and simultaneously control of all microclimate parameters, or simple systems that control, only one parameter.

The harmful substances of the work environment are eliminated from the area by the suction installations. The principles of work are:

- air, charged with harmful substances are sucked through the openings of various shapes and types, located closer to the generation sources;

- sucked air is led through the ventilation channels to a ventilator;

- vitiated air is cleaned with the help of filters, that do not allow evacuation outside, the air that might pollute the environment.

For the chambers (rooms) air refresh, are used the installations that introduce fresh air (see figure 4 [14]). The fresh air is brought from outside, is filtered by dust and as required, is heated, cooled, humidified and introduced into the room through the openings, called discharge holes, placed in the air pipes [3].

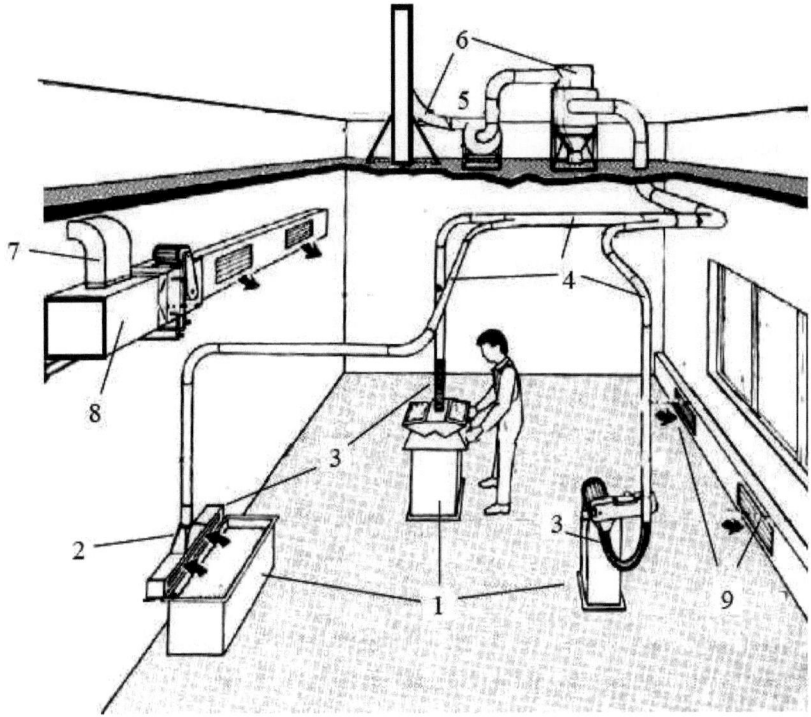

Fig. 4 *Principle scheme of industrial hall with afferent work stations:*
1 – work stations, 2 – pollution source, 3 – pollutant captivation, 4 – transport network of the polluted air, 5 – ventilator, 6 – air vitiated purification, 7 – air captivation system, 8 – general ventilation (fresh air contribution and maybe heated), 9 – general ventilation (vitiated air evacuation)

The intensity of the chamber ventilation by introducing fresh air is measured by the number of air changes per hour, of the respective chamber.

The ventilation systems represent the assemblies of ventilation installations that simultaneously ensuring the microclimate or air purity for a chamber or a chambers group. Independently of the system complexity, its implementation must respect the following general principles:

- the microclimate parameters influence by the ventilation, compulsory requires the air masses circulation between the work zone (internal) and outdoor, air that in some circumstances must be treated and/or recycled;

- the air temperature and humidity control of the work zone is based on mixing this with the circulated air by the ventilation system, possibly heated or cooled, humidified or drying;

- air current speed control is ensured by adequate dimensioning of the air distribution or taking devices, through their correct disposition in relation to areas occupied by operators;

- air purity assurance is achieved either by dilution of vapours or gases of the workspace with clean air, circulated by the ventilation system, until achievement of the maximum permissible concentration (in the case of noxious under gases form and vapours) or by isolating contaminants and their capturing at the production places (in the case of powder, aerosol noxiousness);

- in any situation that requires evacuation from the workspace of the air masses (for capture and removal of the noxious), the maintaining microclimate parameters in the normal range can not be realized only by compensating the exhaust air with adequate treated air.

The classification criteria of the ventilation systems are:
- by the assurance mode of the necessary energy for the air circulation;
- by the action mode of the ventilation in the work area;
- after the pressure regime;
- after the air circulation regime.

By the assurance mode of the necessary energy for the air circulation, are:

- *natural ventilation* – systems in which the air flow necessary energy is obtained from the exploitation of natural phenomena (the action of the climatic factors or disturbances of the workplace microclimate), used to remove excess heat and in some cases, for eliminate noxious lighter than air, if their release is accompanied by heat;

- *mechanical ventilation* – the systems in which the air circulation is obtained by artificial energy utilization, consumed in the equipments which compose the plant in question; the realization of the artificial ventilation involves the using single or in combination of two installation types: of air evacuation or introducing;

By the action mode of the ventilation in the work area, there are:

- *general ventilation* – systems, that assurance the air circulation for the entire space subjected to this action;

- *local ventilation* – systems, that assurance the air circulation in a reduced space;

- *mixed ventilation* – systems, that combine the noxiousness local collecting with the introducing fresh air from the outside, in all ventilated space.

After the pressure regime, are:

- *depression ventilation* – systems that maintain in a room or at the workplace, a constant depression (supply air flow is less than the exhaust);

- *overpressure ventilation* – the systems which are based on the principle of maintaining a constant overpressure in a chamber by introducing and directing the pressurized air in some parts of the working chamber;

- *in balance ventilation* – systems which preserve in a room or at the workplace, the same pressure as the environment, because the supply air flow rate is equal with that exhausted.

After the circulation air regime, there are:

- *ventilation in open circuit (without recirculation)* – systems that realize the air circulation in one direction (air flow rate introduced in the room is full fresh air from the outside – see figure 5 [14]);

- *ventilation with recirculation* – systems in which a part of the exhausted air is used by the air alimentation installation, that is recirculated, in order to saving the energy, consumed for a part of its treatment (see figure 5).

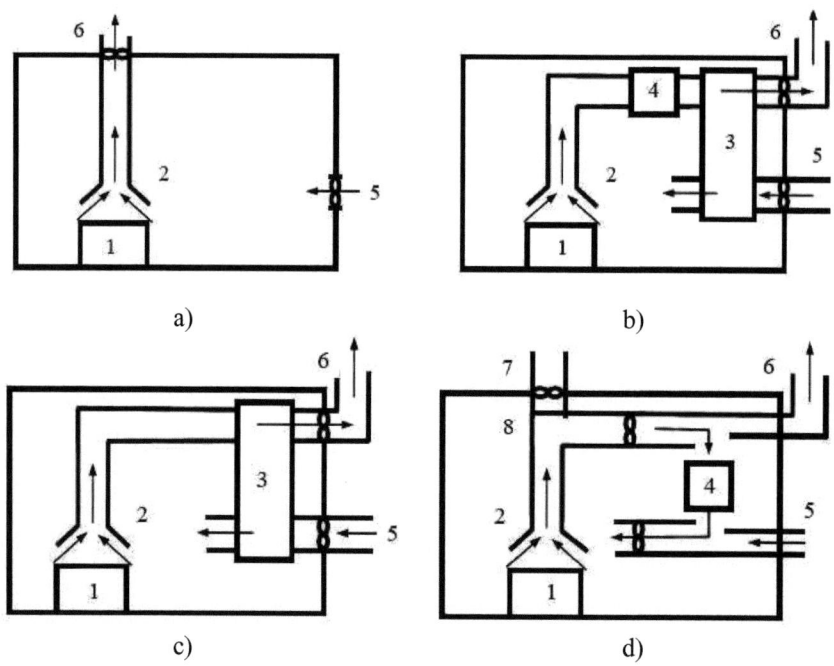

Fig. 5 *Main schemes of the air cleaning and refresh, from industrial spaces:* a) simple evacuation, b) evacuation with purification and recovery, c)evacuation with energy recovery, d) evacuation with recirculation; 1 – work place, 2 – captivation device, 3 – heating recovery, 4 – cleaner scrubber, 5 – air admission, 6 – air evacuation, 7 – safety evacuation, 8 – detour device

The choosing optimal ventilation system depends on primarily by qualitative and quantitative knowledge of the harmful factors existing in the working environment. The quality and quantity of the hazardous substance are determinant factors of the maximum allowable concentration, and its formation and dispersal impose the realization solution of a hygienic environment. Determining the nature and quantity of pollutants, present in the contaminated air is made on the experimentally ways, by analysis and material balance.

Ventilation processes technique – for natural ventilation, air circulation can be achieved by the building leakiness, chaotically – **unorganized natural ventilation,** or by specially arranged openings and adjustable – **organized natural ventilation.**

If a building is exposed to the wind, on the respective side face it is created a pressure, and on the opposite side – a depression. The air enters from the outside into the building, through openings, located on the side exposed to the wind, and will exit through openings, located in the opposite wall. With an adequate positioning of the openings can obtain convenient guidance of the air flow, which produce a rational ventilation of these.

The unorganized natural ventilation is presented in any room, without taking any special measure. Air enters and exits of the room through the walls pores, through leakiness windows, doors, through opening doors and windows. It is considered that in a normal chamber, the air is changed once or twice per hour through natural ventilation unorganized.

Natural ventilation **under gravity effect** can only take place in the rooms where there is heat release. The hot air (heated in workplace area) rises to the top of the room, and its place is taken by fresh air, cold, which penetrating from outside. A prerequisite for achieving such circulation is the existence of openings to the inferior part of the chamber, for the fresh air admission and the openings to the upper part for the warm air evacuation.

The organized natural ventilation requires as the chambers to have special dimensions, shapes and openings. Also without a sufficient height of the room and without heating sources inside, can not be get the organized air flow movement.

In the case of **mechanical ventilation** systems, the main role is of the noxious capture devices, which must be placed in the generation area of the harmful factors, ensuring in this way reducing of the air flow rate, necessary for the chamber ventilation but and increasing the economic efficiency of investment. Due to the large variety of the technique equipments that generate the noxious, in practice, is used, local capture devices of various shapes and dimensions: open, semi-closed or closed.

Whatever of the aspiration type, the capture devices, must fulfil the following conditions:

- to ensure if is possible a complete capture of the emitted noxious;
- not complicate the unfolding and supervision of the production process;
- aspirated air flow rate to be minimum;
- operator to be not found (in normal working position) between the source which generate noxious and the opening of the capture device;
- to aspire only harmful substance particles which tend to spread in the work environment.

In the installations with mechanical ventilation, a particular group is that of the installation dedicated to powder capture and exhaust. The specific feature of this part is the fact that captured powders can not be transported by the piping system only if it provides a minimum airspeed. Otherwise, the powders are deposited, pipes are clogged and the whole installation is compromised. This phenomenon occurs when the system resistance increases and as result, decreases the circulated air flow rate.

The component elements of the ventilation system – regardless of their type and whether or not they are included in a system, the ventilation systems have some common elements. These can be grouped by their role in entrainment, treatment or air distribution, in the following categories:

- elements that provide air movement: ventilators, deflectors (baffles) and skylights;
- elements that provide air treatment: filters, heating or cooling coils, water spray rooms, noxious separators, fan heaters;
- adjustment elements: adjustment flaps, frames with adjustable blinds;
- elements that provide capture, distribution and transport of air: input openings, exhaust openings, air channels (pipes).

The ventilator represents the technical installation which is used to compress air, causing a pressure increase, the principal element being the impeller. The ventilators actuation is realised using electric motors with direct coupling or by v-belts. The ventilators which circulate air loaded with corrosive vapours are made of materials resistant to their action (PVC, stainless steel, bakelite) or other material that adequate protects accordingly. For the air circulation that containing explosive gases must used ventilators of special construction: with aluminium impeller or with increased clearance.

The filters and noxious separators ensure the air purity introduced in ventilated spaces (filters) or the air cleaning, loaded with the captured noxious of the ventilated spaces (noxious separators). They are components of ventilation installations with high resistance to air flow. Most exhaust filters and separators have a variable resistance, given by the fact that during their operation, their resistance increases proportionally with retained noxious volume, so it is necessary them cleaning, regularly.

Heating coils provide the air heating in the ventilation installations. They can work with steam, warm water, hot water or more rare, with electric energy.

Cooling coils are also the heat exchangers of the same construction as the heating coils. By the fins tubes circulate fluid (typically chilled water) at a temperature lower than of the air.

The fan heaters are the heating elements that produce hot air at the place of use. Typically they comprises a sheet metal housing in which is mounted a heating coil and a ventilator driven by an electric motor. They can operate with air of outside, air of indoor or with mixing air.

The spraying water chambers are the heat exchangers, serving to the air treatment by passing it, through a water screen (curtain). Simultaneous with changing the state air, is obtained its purification, by the impurities taking, contained of liquid: dust, smoke, fibres.

The control valves and frames with shutters are the main devices by which to ensure balancing of the different branches resistance for the ventilation systems. With them, it is realised the air flow rate distribution or acquisition from different points, by modifying their internal resistance.

The air channels are designed to transport the air into the ventilation installation, the geometric shape can be circular, rectangular, square, or if required by local circumstances, other cross-section geometry. The air channels must satisfy, the following conditions:

- to be sealed, for as the air taking or evacuation to be made only through the absorption or discharge openings;

- the internal surface of the wall to be as smooth, because the friction pressure loss to be as small as possible;

- to be fire resistant and to have a high thermal resistance, especially when they are used to the warm air.

The materials choice for making air channels must be based on the operating conditions, assembly, construction peculiarities or the economic considerations.

The criterions for selection of ventilation system must correspond largely to workplace needs. For the results to be satisfactory, must take into account the following aspects:

- nature of the harmful discharges;

- amount of the released harmful substances;
- physical and chemical properties of the noxious substances;
- noxiousness degree of the pollutants emitted;
- available space for assembly the ventilation devices and aggregates;
- respecting of the technological requirements;
- satisfaction of some special conditions of comfort for workers.

Some general rules for design of the ventilation systems are:

- to eliminate any generated noxious, the local ventilation through aspiration, is the most efficient method; in this case the devices are ordered in terms of efficacy as follows: casing, semi-closed suction devices, the devices with open suction;

- if it can not apply any of the mentioned devices will appeal to general ventilation;

- in the case of the general ventilation, the suction openings are arranged: at the upper side of the chamber, when the gas or vapours are lighter than air; to the inferior side of the chamber when the harmful escapes are heavier than air; combined, if in the same time are escapes harmful heavier and lighter than the air, or when the specific weight of all substances emitted is not known.

The measurement of ventilation parameters must be checked both at the installation reception and periodically, during operation. For the temperature measurement at the workplace, the following rules will be complied:

- the thermometer should be placed on a stand or suspended by a wire to a height of 1.50 m from the workplace floor; for the places situated to height or in special positions, the temperature is measured to the worker body height;

- the thermometer should be protected during the measurement, of solar radiation or the strong caloric radiation emitted by the equipment, warm and cold air currents, or of the gases;

- the thermometer will be protected against strong air currents with screens.

21

3   Noise combating

The application of solutions to combat noise action is strictly dependent by the correct determination of the risk factor level and physical characteristics.

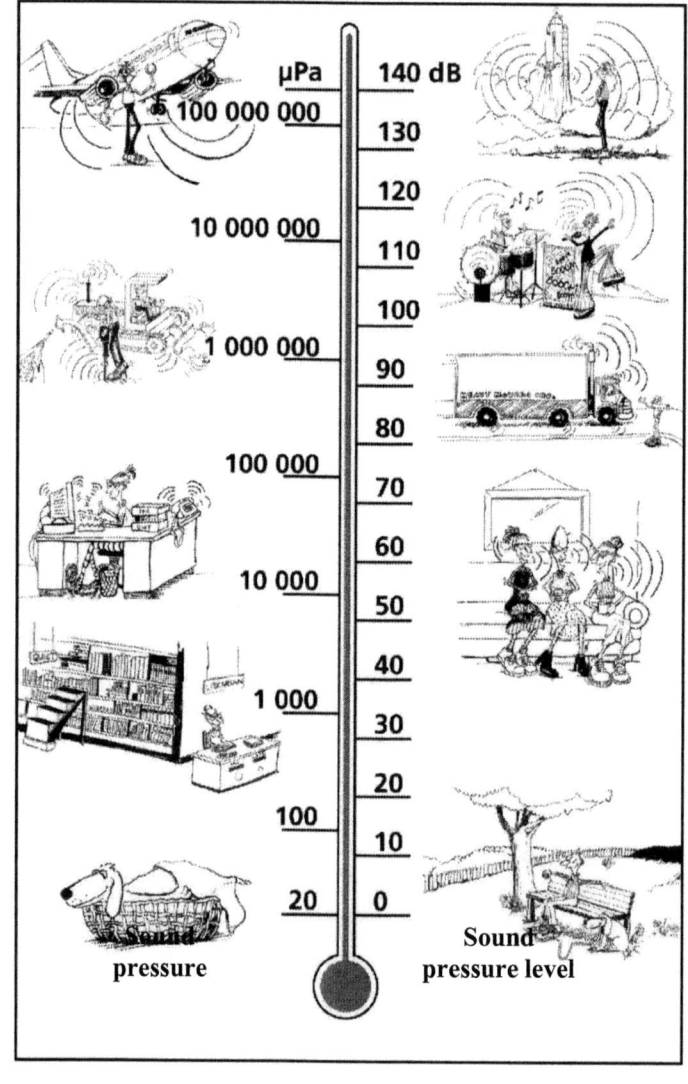

Fig. 6 *Environmental noise level*

For this reason it is important the knowledge of the methods and apparatus that can be used for the noise measurements. Figure 6 shows the noise encountered in various situations of daily life [9].

Methods and instruments for measuring noise – the noise measuring instruments (sound level meters) are complex and must be in accordance with specific standards. The standards contain the necessary information for the measuring instruments, the measurement methods for different types of activities or equipment, the assessment mode of the adverse effects of examined noxious (table 1).

Table 1
Occupational health parameters and associated standards

Parameter	Standard			
	OSHA	**MSHA**	**DOD**	**ISO**
Broadband	A	A	A	A
Broadband peak	Z	Z	Z	C
Exposure time	N/A	N/A	N/A	8
Reference time [hours]	8	8	8	8
Reference time [hours]	40	40	40	40
Threshold level [dB]	80	80	80	70
Criterion level [dB]	90	90	85	85, 90
Peaks over level [dB]	140	140	140	140
Exchange rate for Lav	5	5	4	N/A
Weighting for Lav	S	S	S	N/A

OSHA – Occupational Safety and Health Administration
MSHA – Mine Safety and Health Administration
DOD – Department of Defence
ISO – UK Noise at Work Regulations

The significance of the parameters listed in table 1 is as follows [10]:

- broadband – there are four possibilities for the frequency weighting: A, B, C and Z; A – corresponds to the human ear's response at low to medium sound level (approximately 40 dB); B – corresponds to the human ear's response at medium sound level (approximately 70 dB); it is the most commonly applied frequency weighting and is used for all levels of sound; C – corresponds to the 100 dB equal loudness curve, that is to say, the human ear's response at fairly sound levels; mainly used when assessing peak values of high sound pressure levels; Z does not benefit any frequency weighting, that is equivalent to linear;

- broadband peak – is measured a peak of broadband;
- exposure time is the actual time at that a person is exposed to noise during a workday;
- reference time is needed for calculating the average exposure of the worker and can be extended up to 5 days;
- threshold level is the noise value that makes the distinction between values that contribute or not to the noise dose measurement; those below the threshold are not included in the noise dose calculation;
- criterion level is the maximum averaged sound level allowed for a 8-hour period, when the noise dose is complete (100%);
- peaks over level – any peak levels that exceed the level set here will be counted;
- exchange rate for Lav represents the increase in noise level that corresponds to a doubling of the noise level;
- weighting for Lav represents the time weighted in accordance with this parameter.

The sound level meters may respond differently to the changes sound level, as follows:

- fast – the instrument (apparatus) is able to react quickly, that can track and measure a level that changes quickly;
- slow – the instrument offers a more cushioned and that integrates rapid changes that would make impossible the instrument reading.

In the situations where it is necessary to measure or detect the impulsive noise or peak values, the sound level meters is able to retain the maximum value, thus easing, the results reading.

The *A* network weighting, allows measurement of subjective sensation of auditory perception. How the human ear is less sensitive to low frequency sounds, *A* weighting consists primarily in a low and medium frequencies reduction. Using the *A* weighting network, permits to measure the weighted global noise level L_A, in dBA, and the assessment of the continuous equivalent noise level L_{Aeq}, in dBA. L_{Aeq} is a standardized form of an average noise level on the long term. A person exposed to the same total acoustic energy in a period of time, either that is exposed to the effectively noise intensity, including all of its fluctuation, or to the constant intensity L_{eq} of the noise exposure for the same period of time.

The concept is extremely useful when it comes to typical industrial noise that fluctuates a lot and contains different periods of intense or less intense noise. In industry, the permissible noise exposure for the employees, could thus be defined as the maximum value of L_{eq} for a normal working period.

The ISO R 1996 and ISO R 1999 standards, requires a half-life of the exposure time every 3 dB increase in noise. OSHA (Occupational Safety and Health Administration) requires a half-life of the exposure time, to the increased noise levels by 5 dB.

Another parameter provided by the sound level meter, is the instantaneous noise L_{AF} (A - weighting network A, F - fast mode).

General solutions for noise control - the main aim in most cases is to reduce the risk of developing occupational deafness. Another objective is to ensure protection in terms of general effects, that is avoiding the installation of embarrassment and interference effect of the attention, in the context of providing communications intelligibility (especially safety signals) is very important. It also seeks the extra effects reduction of the noise.

The figure 7 shows statistical data for the 2007 year, provided by the Statistical Office of the European Union (Eurostat) and describes the persons reporting the physical factors they were most exposed to by type. This figure shows that in all analyzed countries, even at European Union (27 countries) level, the noise is one of the present risk factors (the 4th place). The conclusion, the workplace noise level must reduce on any ways, for a better comfort of the employees [7, 15].

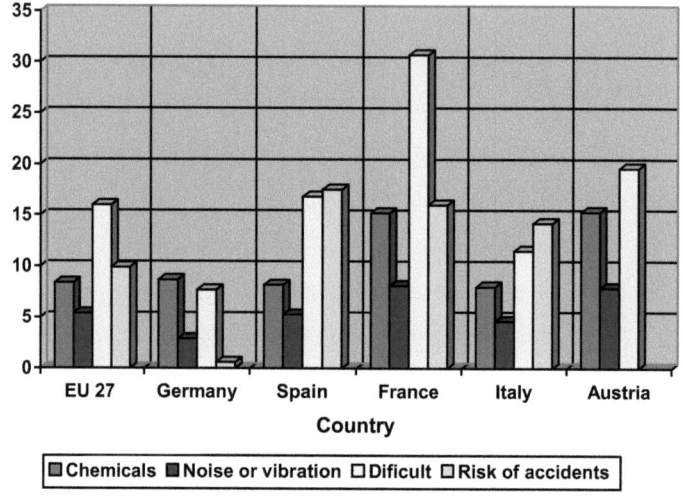

Fig. 7 *Main physical factors of perturbation for persons, at workplace*

Chemicals – chemicals, dusts, fumes, smoke or gases
Difficult – difficult work postures, work movements or handling of heavy loads

The figure 8 shows statistical data for the 2007 year, provided by the Statistical Office of the European Union (Eurostat), and describes the persons reporting the exposure to factors than can adversely affect physical well-being, by economic activity sector. Analyzing this figure is obtained, that indifferent of country the sectors order with problems at occupational health level is almost same [7, 15]:

- mining and quarrying;
- construction;
- manufacturing;
- transport, storage and communication;
- hotels and restaurants.

26

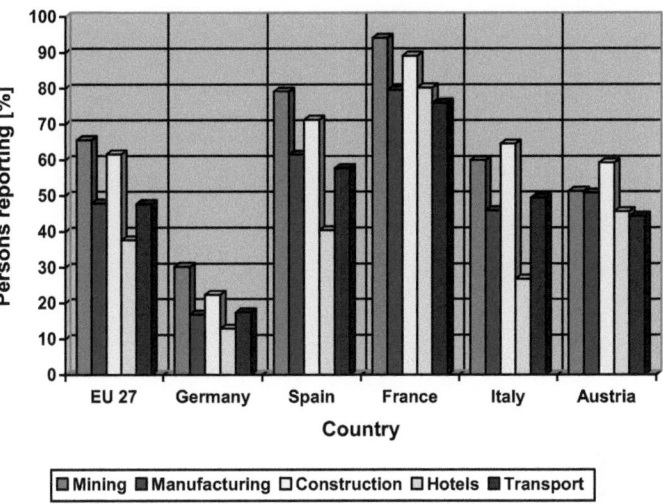

Fig. 8 *Evolution of the exposure to risk factors by activity sector*

The mining and construction sectors record the most exposure of the risk factors, reason for that, in these sectors, the quality of the workplace must be much improved.

A first objective of the noise combating solutions is to reduce the risk to appearance of the occupational deafness. Also is desired and the diminution of the extra auditory effects of the noise. The noise combating is a system problem, through this understanding the assembly formed by the noise sources, environment (ways) of the acoustic energy propagation and the receiver. The noise combating methods belong to the described above system and should be part of the following categories:

- combating methods at the noise source;
- noise combating methods to prevent propagation;
- noise combating methods at the receiver (see Appendage 1).

The solutions used to the noise combating are:

- noise attenuators;

- sound absorbing casings;
- sound absorbing screens;
- sound absorbing cabins.

The noise attenuators are part of the noise combating methods at source, constituting an effective and frequent solution used, to reduce the aerodynamic nature sources. Depending by the construction, attenuators can be:

- active – are constructed as a channel shape lined with sound-absorbing material;

- reactive – ensure the acoustic energy dispersion by forming a waves barrier, that prevents the sound passing at some frequencies, because of the air mass and elasticity influence, in the attenuator cells.

The reactive attenuator is composed by cavities and tubing series, mounted in such way that by the created discontinuities, to realize the required noise reduction.

The sound absorbing casings are one of the most common methods of noise reduction on the propagation ways, consisting in the complete closing of the noise source in sound absorbing precincts. In the case of the sound absorbing precincts, noise can propagate to outside in the following ways:

- through the casing walls (air noise);
- through leaks (leakiness) or the technological openings (air noise);
- through casing structure (structural noise);
- through machine-tools elements (technological equipment).

The attenuation of the air noise level, transmitted through sound absorbing casing walls, will be determined by the relation 3:

$$\Delta L = R - 10 \cdot \lg \frac{S_c}{A_c} \quad \text{[dB]}, \tag{3}$$

where:

R – noise reduction of the casing wall [dB];

S_c – the wall casing surface [m^2];

A_c – total acoustic absorption, of the casing walls interior surface (see the relationship 4).

$$A_c = \sum_i \alpha_i \cdot S_i \ , \tag{4}$$

where:

α_i – sound absorption coefficient for the i casing element;

S_i – interior surface of the i casing element.

For the particular case in that entire inner surface of the casing is formed of the same sound-absorbing material, the relationship 3 becomes 5:

$$\Delta L = R - 10 \cdot \lg \frac{1}{\alpha} \ \ [\text{dB}]. \tag{5}$$

In conclusion, we can say that the attenuation realized by a sound absorbing casing, depends by the R noise reduction of the casing wall and by the sound absorbing material quality used in its manufacture. Based on the mass law, the noise reduction can be determined by the relation 6:

$$R = 20 \cdot \lg (p \cdot f) - 47,5 \ \ [\text{dB}], \tag{6}$$

where:

p – weight of a square meter (of surface) of sound-absorbing element (dividing);

f – frequency [Hz].

Because in most cases, the casing walls have not a homogeneous structure, their continuity being interrupted by assembling of the various elements that compose them (doors, viewing windows, channels equipped with the noise attenuators) that worsen the noise reduction, the total amount of this, for the wall of, this casing type is determined by the relationship 7:

$$\overline{R} = R - 10 \cdot \lg\left[1 + \frac{S_i}{S_0} \cdot \left(10^{0,1 \cdot (R - R_i)}\right) - 1\right] \quad [dB], \tag{7}$$

in which:

    R – sound reduction of the wall, without openings [dB];

    R_i – noise reduction of the various elements [dB];

    S_0 – total area of the wall, including openings [m²];

    S_i – opening surface [m²].

In order to reduce the structural noise transmitted to the outside of the casing, must as all connections between the machine tool elements and the casing respectively the floor, to be realized in a vibro-attenuation (insulated) construction.

The sound absorbing screens, known as sound absorbing panels (see fig. 9) represents acoustic barriers placed between the noise source and receiver to noise combating, on the propagation ways of its.

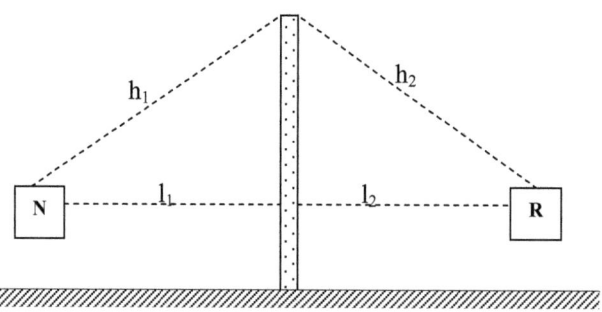

Fig. 9 *Sound absorbing screen placing draft*
(N – noise source, R – receiver)

Considering a noise source N and an observer R, between which is a sound absorbing screen (see fig. 9), the attenuation realized by this, is calculated using one of the relations 8, 9 and 10:

- for frequencies below 500 Hz:

$$\Delta L = -20 \cdot \lg \frac{l_1 \cdot l_2}{h_1 \cdot h_2} \quad [\text{dB}]; \tag{8}$$

- for frequencies between 500 and 1,000 Hz:

$$\Delta L = -20 \cdot \lg \frac{l_1 \cdot l_2}{h_1 \cdot h_2} + 2,5 \cdot \left(\frac{f}{500} - 1 \right) \quad [\text{dB}]; \tag{9}$$

- for frequencies above 1,000 Hz:

$$\Delta L = -20 \cdot \lg \frac{l_1 \cdot l_2}{h_1 \cdot h_2} + 3,5 \cdot \left(\frac{f}{1000} - 1 \right) \quad [\text{dB}] . \tag{10}$$

The notations used in equations (8, 9 and 10) are:

l_1 – distance from the source to the screen;

l_2 – distance from the screen to the observer;

S_0 – screen surface;

r_0 – radius of the circle whose area is equal to S_0.

$$h_1 = \sqrt{l_1^2 + r_0^2}; \quad h_2 = \sqrt{l_2^2 + r_0^2}; \quad r_0 = \sqrt{\frac{S_0}{\pi}} . \tag{11}$$

h_1, h_2 and r_0 are calculated according to the relations 11. It is recommended as the width of the screen to be greater than the height of at least

1.5 – 2 times. The sound absorbing screens efficiency depends of their dimensions, of acoustic frequency of the source, and in the majority cases the soundproofing is not realized to the desired parameters by applying a single solution.

Sound (phono) absorbing cabins (boxes) can be used to combating noise only in the case of processes that can be commanded, controlled and/or distance monitored, minimizing the workers exposure time, to this risk factor. To the sound absorbing cabins realization, must respected the following rules:

- materials used to cabins construction must ensure the air noise isolation capability, required by the necessary attenuation characteristic;

- doors and windows are made in sound insulation construction;

- doors and windows sealing systems, must not diminish the cabin attenuation characteristics;

- cabin equipping with noise attenuation for the ventilation or circulation air system;

- coupling of the sound absorbing cabin with the building structure, must be realized through a vibro-insulator system.

Noise effects on the work environment – protecting workers from the effects of noise can be a difficult task because must taken into account the particular characteristics of noise in the workplace and other existing risks in the workplace [7]. The noise adverse effects on the human body are emphasized primarily through, hearing deficiency or ears lesions, with the concomitant exposure to oto-toxic substances. But the noise exposure can "cost" more than hearing loss. The adverse effects of noise are: the increased risk of work accidents, the foetus injury during pregnancy (pregnant workers) and stress, which in turn brings other drawbacks (troubles). An undesirable effect, but important of the noise is affecting of the cardiovascular system health.

The noise causes an adrenaline discharge, which acting on the blood vessels, determines high blood pressure, ischemic heart disease and the vascular cerebral accidents. The most important death cause is of the cardiovascular disease whose appearance and evolution is adversely affected by the noise intensity and duration. By increasing the number of illnesses, the expenses for society but and for the each person becomes more and more hard to supported.

According to studies, one of four workers in the European Union suffer from stress at work, which leads to its level, to costs of billions Euro in lost working time, health care and other expenses. In generally, stress at work is determined by many factors, one of these being the noise. The noise must not be strong to cause a stress state. For example, sometimes it is enough a weak noise, but persistent. How the noise affects the workers concentration ability, depends on the following factors:

- nature of the noise, including volume, tone and its predictability;
- work task complexity effectuated by the worker;
- fatigue state and physical form of each person.

In Europe is recorded annually over 7.5 million accidents at work. Although it is difficult to quantify the noise role in the production of these accidents, the logic and the dates indicate that this role can be significant. The noise can lead to accidents by:

- verbal communication disruption between workers;
- sound masking of the imminent danger or of warning signals;
- distraction of workers attention;
- work stress increasing.

The noise causes a decrease of the work efficiency in the activities that require a continuous and enlarged attention if are perceived unusual sounds for the workplace, unexpected or discontinuous and the high frequency sounds. In the case of monotonous activities the noise can improve work efficiency. The strongest disturbances caused by noise, appear in the activities that require attention and maximum concentration. The studies show that the noise influence on the work efficiency, increases if people perceive the noise to they are exposed as being annoying or unpleasant. Many people perceive the absence of noise as a convenience.

According to **European Union Directive 2003/10/EC** are set following values [13]:

- **exposure limit value – 87 dB;**
- **lower exposure action value – 80 dB;**
- **upper exposure action value - 85 dB.**

4   Vibration combating

The human vibrations are defined as an effect of the mechanical vibrations produced in the environment, where the human body is. In everyday life, the people are exposed to various sources of vibration, caused by the bus, train or car travelling. Also, many people are exposed during a working day to a different kind of vibration, produced this time, by the work equipment: hand tools (drills), machine tools and heavy vehicles (vehicles earthmoving) [11].

The vibrations generated by the moving parts of work equipment, vibration that could not be removed by the initial design, through the suitable dimensions choice or measures taken to source, are transmitted to their fixed parts and then through links between work equipment and building, they will be forwarded in the form of elastic waves, to the building elements [4, 5]. For as the vibration action to do not constitute a danger, is necessary as these forces to be as small.

Basically the combating of vibration action is made by isolation and/or vibration damping. A concept of particular importance in the study of the vibration isolation and damping is that of transmissibility. This allows to determine which is the dynamic force generated by the vibration source and transmitted to the surrounding elements. By the transmissibility analysis, can be determined the assembly, which must elected for as, the transmitted force to be as small as possible.

In the case of static work equipments, for the design an anti-vibration isolation, that to serve them, have taken the following steps:

- knowledge of the working equipment: dimensions, power, work rotations, kinematic scheme, size, direction, frequency and points of the disturbing forces application, gravity centre's position, the pipes placing and other accessories;

- knowledge of the buildings, equipment and neighbouring installations, indicating their vibration sensitivity;

- calculus model selection and its parameters: the anti-vibration isolation can be obtained either by the placing elastic insulators between the work equipment and work floor (for small work equipment), or fixing the work equipment (WE) on a reinforced concrete block, elastic leaned;

- vibration amplitude calculus: is determined the WE amplitude movements, like those of the forces transmitted to the ground or floor;

- resistance design based on the known static loads and of those dynamic, result by the anti-vibration isolation calculus;

- measuring of elastic constants and WE insulators, the damping measuring, systematic measurement of WE vibration.

In conclusion, to anti-vibration isolation achievement are necessary two building elements, known as the vibro-insulator and vibration damping.

Vibro-insulators are the elastic elements, which through their deformation make possible the vibratory movement, respectively the anti-vibration insulation. After the exterior appearance, the elastic elements under the work equipment can be divided into two groups:

- plates – usually placed under the entire assembly (equipment), made of cork, felt, rubber, pressed fabrics, masonry, reinforced concrete or foundation soil itself;

- discrete elements – made of steel spring, rubber, synthetic materials, poles of steel or reinforced concrete.

Cork is used as plates with usual thicknesses of 60 mm, either of natural cork or of cork granules and pressed, soaked with binder. Natural cork is elastic, water and oil resistant, retaining its elasticity over 40 years. The main applications of cork plates are to the working equipments with high rotations (electric motors, turbo generators) in the places where it is necessary an elastic surface and where the steel springs do not offer an adequate solution.

To avoid the cork fragmentation, it is recommended as the static pressure does not exceed 0.20 MPa (2 daN/cm^2).

Felt is a material more resistant to chemical and mechanical agents than cork and rubber. It currently uses as the elastic layer and the noise absorber. It can accept admissible pressures of 6-8 MPa (60-80daN/cm^2).

Pressed fabrics are made of the jute fabric superposed, impregnated against humidity, pressed so as, to form the resistant plates. These plates have the advantage of the lasting longer elasticity keeping and are less sensitive than felt to the corrosive agents. Plates can resist to high pressures, up to 30 MPa (300 daN/cm^2).

Rubber is a material widespread in the vibro-insulators construction, because of the properties that make it indicated for this:

- reduced elasticity modulus to compression, so that the rubber is very much deformed, being able to take by shock, a mechanical working over 4 times greater than that taken by the steel springs;

- small elastic constant, so that the machine tools pulsations placed on it, are reduced;

- high damping, being possible up to 30-50% dissipation of the vibration total energy;

- to the equal efficacy, occupy, relative to other material, a reduced area, has a low weight and can be put to work together with steel springs.

Also the rubber, has some disadvantages to vibro-insulators realization:

- its elastic properties loses under the influence of atmospheric agents and especially liquids (water, oil, acids);

- in time, after 5-20 years old, presents the ageing effect, losing of the elastic properties, in this mode, endangering the suspension system.

To implement the rubber vibro-insulators, it is used, both, the natural and synthetic rubber, these can be subjected to tensile, compressive efforts, shear and torsion efforts.

Steel springs are one of the best solutions of isolation against vibrations. Due to large deformations, they allow the realization of vibro-insulation suspension with very lower own frequencies. Unlike the rubber vibro-insulators used to small and medium forces, the steel springs can be constructed for the various tasks, respectively, from the lightweight machinery to heavy machinery. After their constructive shapes, are distinguished: helical springs, torsion bar springs, leaf springs, ring and disc springs. After the external strength are distinguished: tension springs, compression, bending and torsion.

Dampers are elements used where the resonance amplitude is large, being able to quickly eliminate the vibration or shock energy. After the operating principle are: friction dampers with fluid layer (hydraulic dampers), with Columbian friction dampers and with Eddy current losses dampers. Among these types of dampers, the most used is the group of the telescopic dampers.

Methods and tools for vibration measuring – to determine the vibrations to which the human body is exposed during work time, are used the apparatus of the human vibration analyzers type (vibrometers).

There are two main types of human vibration:

- of the whole body;
- of the hand-arm.

In the first case, the vibrations are transmitted to the whole body, the affected parts being: feet, back and buttocks. Prolonged exposure to this type of vibration can cause permanent physical damage or disturb the nervous system. In the second category, daily exposure to hand-arm vibration over a number of years, can cause permanent physical damage, usually resulting in what is commonly known as "white-finger syndrome", or it can cause damage the joints and muscles of the wrist and/or elbow. Much research and many studies have been made to evaluate the effect of over-exposure to human vibration, especially in working environments. The results have been used to establish international standards that allow human exposure to vibration to be evaluated.

For the workers protection, when they are exposed to risks arising from vibration in the course of their work, was emitted the European Union Directive 2002/44/EC that introduced minimum Health and Safety requirements for these. Also, the directive requirements must corroborate with the next standards:

- **ISO 8041:2005 – Human Response to Vibration – Measuring instrumentation;**

- **ISO 20643:2004 – Mechanical Vibration – Hand-held or hand-guidance machinery;**

- **ISO 5349-1 – Mechanical Vibration – Measurement and evaluation of human exposure to hand-transmitted vibration – part 1: general requirements;**

- **ISO 5349-1 – Mechanical Vibration – Measurement and evaluation of human exposure to hand-transmitted vibration – part 2: practical guidance for measurement at the workplace;**

- **ISO 2631-1 – Mechanical Vibration and Shock – Evaluation of human exposure to whole-body vibration – part 1: general requirements;**

- **EN 14253 – Mechanical Vibration – Measurement and calculation of occupational exposure to whole-body vibration with reference to health – practical guidance.**

The parameters that can be determined with vibrometers help are:

- **elapsed time (ET)** – measurement duration, counted from start of the measurement to stop of the measurement and excluding the time when the instrument is paused;

- **running r. m. s. acceleration value (Curr RMS)** – the frequency-weighted running r. m. s. vibration acceleration value is measured using exponential averaging with a time-constant of 1 second; this is an instantaneous value displayed during measurement;

- **maximum transient vibration value (MTVV)** – the maximum value of the running r. m. s. acceleration value measured over the elapsed time: the time constant is equal to 1 second;

- **time-averaged weighted acceleration value (Total RMS)** – the frequency-weighted running r. m. s. is measured using linear averaging with an averaging time that is equal to elapsed time;

- **peak vibration value (Peak)** – the maximum modulus at the instantaneous (positive and negative) peak values of the frequency-weighted acceleration measured over the elapsed time;

- **vibration total value (VTV)** – the combined vibration from the three axes defined as the square root of the sum of the square of the vibration values multiplied by the k-factors for the three axes (see relation 16); the k-factors are multiplying factors that depend on whether hand-arm or whole-body is measured;

- **vibration dose value (VDV)** – the fourth root of the time integral of the fourth power of the instantaneous frequency-weighted vibration acceleration (see relation 17); measurement in $m/s^{1.75}$, the integration time is the elapsed time;

- **8-hour vibration dose value (VDV(8)k)** – the VDV measured over elapsed time is extrapolated/interpolated to the value that the same signal would have given if elapsed time was 8 hours and multiplied by the respective k-factors.

A(1), A(4) and A(8) represent the daily vibration exposure value for 1 hour, 4 hours and 8 hours exposure to vibration. The Vibration Directive (Directive 2002/44/EC) sets minimum standards for controlling the risks, both from hand-arm and whole-body vibration.

For hand-arm vibration, the directive sets an exposure action value, above which it requires employers to investigate the hand-arm vibration risks on their workforce and an exposure limit value above which employees should not be exposed. These are:

- a daily exposure action value of 2.5 m/s^2;
- a daily exposure limit value of 5 m/s^2.

Fig. 10 *Anti vibration gloves*

For the assessment of the daily vibration exposure, an estimate of the time that the tool operators are exposed to the vibration is required. In practice, two main methods of daily exposure calculation are used, these depending by the operation type, which can be continuous or intermittent. In the case of the intermittently operations, only the contact time should be recorded. This is the time that the hands are actually exposed to the vibration from the tool or workspace. The hand-arm vibration is attenuated by the anti vibration gloves (see fig. 10).

The calculus relation of the daily vibration exposure is expressed as 12:

$$a_{hw} = \sqrt{\frac{1}{T} \cdot \sum_{j=1}^{N} a_{hwj}^2 \cdot t_j} \quad , \tag{12}$$

where:

a_{hwj} – measured vibration magnitude (Total RMS);

t_j – measurement duration (ET) of sample j;

N – number of partial times.

The total exposure time T is the sum of all partial exposures and is calculated with relation 13:

$$T = \sum_{1}^{N} t_j \quad . \tag{13}$$

According to the ISO and ISO 5349-2 standards, the total measuring time should be at least 1 minute and at least three measurements should be completed for each operation. For the intermittent vibration case, the relations 14 and 15 permit for the A(8) and a_{hv} values calculus:

$$A(8) = a_{hv} \cdot \sqrt{\frac{T}{T_0}} \quad , \tag{14}$$

where:

T_0 – 8 hours;

T – effective exposure time (T=8, 4 and 1 hour).

$$a_{hv} = \sqrt{a_{hwx}^2 + a_{hwy}^2 + a_{hwz}^2} \quad . \tag{15}$$

were:

a_{hv} – weighted energetic sum of X, Y and Z.

The whole-body vibration is applicable when the vibration motion is transmitted to the human body through a supporting surface. Usually, the persons exposed to this vibration type, are: farmers, operators of the extraction industry equipments, or earthmoving equipments, truck drivers, navigators and helicopters pilots. For whole-body vibration, the Vibration Directive (Directive 2002/44/EC) sets the next action values, above which it requires employers to investigate the exposure vibration risks:

- 0.5 m/s^2 for the daily exposure action value (or at the choice of the individual EU Member State, a vibration dose value of 9.1 m/s$^{1.75}$);

- 1.15 m/s^2 for the daily exposure limit value (or at the choice of the individual EU Member State, a vibration dose value of 21 m/s$^{1.75}$).

A(8) value, for the whole-body vibration, is calculated with relation 16:

$$a_{wv} = \sqrt{k_x^2 \cdot a_{wx}^2 + k_y^2 \cdot a_{wy}^2 + k_z^2 \cdot a_{wz}^2} \quad , \tag{16}$$

where:

a_{wx}, a_{wy} and a_{wz} – vibration values on the three orthogonal axes X, Y and Z;

k_x, k_y and k_z – multiplying constants whose values depend on the measurement application.

The vibration dose value (VDV), which is an alternative indicator of vibration exposure, is calculated by the relation 17:

$$VDV = \left\{ \int_0^T [aw(t)]^4 \cdot dt \right\}^{1/4} , \tag{17}$$

where:

     T – total (daily) period for which exposure occurs;

     aw(t) – instantaneous frequency-weighted vibration acceleration.

5   Workplace lighting

By lighting conditions for workplace, they are understood the values received of the various parameters that characterizing the light environment provided by a lighting installation. The objectives of the lighting installations are:

- providing in the work area of a light quantity, capable to ensure fast and accurate perception of the elements that constitute, the visual task;

- distribution of the light, so as to achieve a good balance of luminance in the room and work area;

- correct colour rendering (given);

- maximum efficiency ensuring in terms of investment and operating costs (maintenance, electricity energy).

To performing lighting installations must respect the following standards: **ISO 8995-1:2002 and EN 12464-1:2002.**

Lighting conditions for the working spaces – the main parameters that characterize the bright ambience of a work area, are:

- average illumination on the plane or on work surface;

- uniformity of illumination;

- reflection coefficients of the main surfaces (ceiling, walls, work area, floor, equipments, facilities);

- proportion between the work area luminance and the general visual field;

- ratio between the main areas illuminations that make up the structure of the visual field;

- protection angle of the general lamps and local lamps;
- work surfaces aspect;
- pulsatory character of the light that reaches the work plane;
- colours rendering index for the light sources used;
- the lighting direction.

The lighting level – depending of the size details, of the contrast between details and background, but and the brightness and background, occupational health and safety standards establish lighting levels, that must be provided, for indoor work spaces, according to the visual working categories, of realized.

Uniformity of lighting – general and local lighting installations can not provide a constant illumination on the work surface or general plan. In order, to do not disturb the vision process, the illumination variation must remain within certain limits.

Uniformity of illumination is expressed by C_u factor, which is the ratio between the minimum illumination (E_{min}) and average illumination (E_{med}) or maximum (E_{max}) on the considered surface:

- $C_u = E_{min} / E_{med}$ – in general plan case;
- $C_u = E'_{min} / E'_{med}$ – on the working surface of the table, machine, panel.

The value of uniformity factor depends of the work fineness and by the reference surface. The visual adaptation effect decreases at the direction changing from one point to another, on work surface or the general vision field, when difference between the illuminations is less. The illumination values E_{min}, E'_{min} and E_{max} are measured and E_{med} is determined in accordance with the standards [4, 5, 6].

If the work process imposes, the personnel frequent movement in different locations, or between these and the corridors, stairs, the average illumination ratio of neighbouring spaces, must be $E_{med1} / E_{med2} \geq 0.1$.

Ambience light quality – lighting levels established according to the standards, ensure visual efficiency as expected, when lighting conditions quality satisfy the following aspects:

- avoiding or limiting direct and reflected glare;
- more accurate colour rendering;
- removal or attenuation of the pulsatory aspects of the emitted light and stroboscopic effect when using gas discharge lamps;
- adequate modelling of the general visual field structure.

Avoiding or limiting direct and reflected glare – lighting quality, is assessed by luminance value that are obtained in the lighting installation, on each of surfaces that compose the visual field structure. The presence, in the visual field of a surface with greater luminance, than the luminance to which the eyes are adapted, can produce glare, which has the effect, of reducing visual efficiency and visual comfort. All the artificial light sources, sun, light colours surfaces illuminated strong and their image on the glowing surfaces, can produce the same effect.

To avoid or limit the direct and reflected glare, the next are followed:

- finishing of the chambers main surfaces;
- luminance ratio on the work surfaces;
- ratio between main surfaces illuminations that delimit the general visual field (ceiling, walls, work surface, equipment, facilities);
- protection angle of the lamps;
- aspect of the working surfaces.

By the ratio limiting, between the main surfaces illumination, that compose the structure of the general visual field is desired, assurance of the necessary conditions for the eye relaxation, without excessive effort of adaptation and visual fatigue. The ratio value between detail luminance and background luminance on that, this is seen, is of 1/3 and in the case between the detail luminance and the surfaces luminance, immediately surrounding the background, this ratio is 1/10.

The ratio values were set so that, considering the psycho-physiological effects of blindness, to be ensured three levels of quality for lighting installations, in function by the work difficulty categories:

- *without glare,* category I;
- *scarcely perceptible glare* for categories II and III;
- *acceptable blindness* for categories IV and V.

The light sources from lamps and windows are usually sources of direct glare and visual discomfort. For this reason, in rooms where are made the exacting work (visual categories I-IV), are imposed the following measures: the protection or sources mounting, in off the angle 45° measured from the horizontal line of view, considering the normal working position and total or partial coverage of the windows with curtains.

If the surfaces are polished, there is a risk of apparition of glare by reflection of the light sources on these surfaces. To avoid this, will be practiced: mat finishes, a position corresponding to the lamps so that their image is not visible from the working position and the light sources protection, in work surface direction with screens, of white diffusing material.

More accurate colour rendering – it results of the surfaces colours interaction of the light area, causing physiological and psychology reactions for operator. It is used as a factor to improve the workers' psychological climate, lighting and even noise, at work. To establishment the chromatic environment must take into account the influence of the colour combinations on the operator sensations, of the image perception and of the text legibility from distance.

Colour affects the size and position apparent of the objects. The finished in warm colours, intense, seem closer, while cool colours seem more distant. The intense colours attract attention, warn, irritating and those diluted give the feeling of calm and tranquillity. Considering some ambience factors, specific to the working chambers, such as temperature, spatiality, noise, illumination level, it can be determined the colour scheme, that to improve their characteristics of point of view of the occupational safety and health (see table 2).

Table 2

Colours combination influence on operator sensations

Colours combination	Produced sensation		
	Pleasant	Doubtful	Unpleasant
Red with:	navy blue, green	yellow	violet-purple
Orange with:	blue, green, violet	red	yellow
Yellow with :	purple, blue	red-violet	green, orange
Green with:	red, violet	purple	blue, orange
Violet with:	green, orange	yellow	red, purple

Typically, in an ambience scheme is selected one or two dominant colours, all others will be harmonized with them. Table 3 presents the order to distance legibility of the various colour combinations.

Table 3

Distance legibility order of different colour combinations

No. of order	Colours combination	No. of order	Colours combination
1	Red on yellow background	7	Blue on white background
2	Yellow on black background	8	White on black background
3	Green on white background	9	White on blue background
4	White on red background	10	Red on black background
5	Red on white background	11	Black on white background
6	White on green background	12	Green on red background

If are respected the chromatic ambience basic rules, warm colour schemes should be applied, stimulants:

- in rooms exposed to the north;
- in rooms where there are high noise sources;
- in places where the process requires low temperatures;
- if the room is too large high and has a flat aspect;

47

- in rooms where physical effort is low, the exposure times of visual tasks are short and is imposed a stimulating atmosphere.

Cool colour schemes, restful, soothing, it is recommended to:
- rooms exposed to the south;
- spaces where are exothermic processes;
- spaces where the noise level is high;
- situations which aims to give an appearance of spaciousness to small spaces and with a crowded structure of the visual field;
- spaces that requires a calm and restful atmosphere.

In terms of the intensity of the colours that compose a scheme, it will be inversely proportional to the surfaces size, on that respective colour is applied. It will also take into account the possible role of colour: warning, ban, attention, signalling, as set of occupational safety norms.

The colour contrast distinguishes the structure of the visual field, in same time maintaining a uniform distribution of luminance. To highlight areas can use different colours well harmonized, that have close values of the reflection coefficients. The primary colours using, in addition to mental and visual ambience, ensure the obtaining in economical mode of a uniform and correct lighting, owing to the multiple reflections, that occur at the light colours on large surfaces of workrooms.

Removal or attenuation of the pulsatory aspects of the emitted light and stroboscopic effect when using gas discharge lamps – the mercury vapour lamps, sodium vapour lamps and those tubular fluorescent are characterized by a periodic variable emission of the bright flux, owing to the arc discharge current because the supply is with alternating voltage.

The current frequency is 100 Hz, except the tubular fluorescent ends where 50 Hz is. Frequency of 100 Hz is large enough to be perceived visually as a continuous emission. Frequency of 50 Hz is detected by the peripheral areas of the retina, when the tubular fluorescent lamps ends are seen with tail of the eye.

The effect of light emission variation is that it produces false image of the real motion for the objects that rotate or have a linear motion with certain frequency, as well as visual fatigue.

Therefore, in spaces in which are realized the precision visual works and in places where there are work equipments with moving parts that can give the stroboscopic effect, for lighting flux pulse attenuation on working surface, it can be applied diminishing measures, such as:

- balanced distribution of light sources on the three phases of the power supply network;

- covering the ends of the fluorescent lamps that are used in local lighting equipment, to a distance of 5 cm.

Adequate modelling of the general visual field structure – the visual field aspect is good when people and objects that enter into its structure are illuminated so, that they can be seen clearly as the shape and location. This is achieved when there is a suitable ratio between the directed light and diffused light.

Directional light is when this arrives prevalent on an object of a single direction. It is put in evidence when is used the light sources of small dimensions but with high power, in rooms with dark colour surfaces.

Diffuse light coming from all directions and is given when are used the large dimensions sources, in rooms with light colour surfaces.

The predominance of the directional light produces harsh shadows, dark, which loads the structure of the visual field, reduces visibility of the details or objects and gives an unpleasant aspect of the workspace. The predominance of diffuse light also is not convenient, as it reduces almost completely shadows, which diminishes the clarity of forms and their position in space. In practical terms, we obtain a suitable distribution of the shadows if the main surfaces of rooms are finished in light colours and general lighting system ensures the lighting ratios, specified in the norms.

Means and solutions to achieve the lighting – light sources, lamps and lighting systems are means by which the lighting is realized in accordance with the each workspace, visual requirement. The lighting can be:

- **natural** – when the light source is solar radiation, radiation that penetrates in the working rooms by gaps practiced, in the building elements;

- **artificial** – when is realized, usually with light sources supplied by powered electricity network, of low voltage.

In the artificial lighting case, one or more light sources are mounted in a lamp that, depending on optical devices that an equips, redistributes the luminous flux emitted by light sources. Depending on the specific visual request, the lamps are distributed uniformly or grouped at the ceiling level or its vicinity, forming the general lighting system of the room or workspace. The local light, located near of the working surface, can supplement this system configuration.

For the workspaces lighting case, can be used incandescent or discharge lamps. These sources considerably differ as physical dimensions, electrical characteristics, spectral distribution and luminous performances. To ensuring the efficiency visual, of the aesthetics lighting installation and the aesthetics industrial design, for the light sources choice, are taken into account:

- all the light sources of uniform general lighting will be of the same type and will have the same spectral distribution;

- in the installations case of located general lighting, will be applied the previous rule; for the single block hall with large area, it is permissible to use fluorescent lamps for the areas illuminate with exacting working, and mercury vapour lamps to illuminate other areas;

- if for general lighting installation, the tubular fluorescent lamps are used, the local lighting installations can use incandescent or fluorescent tube with the same spectral distribution as, of the lamps used for general lighting;

- when for general lighting installations are used the high-pressure mercury vapour with fluorescent balloon, for local lighting installations can be used incandescent or fluorescent tube;

- when for general lighting installation are used sodium vapour lamps, of high pressure, the local lighting installations can use incandescent or fluorescent tubular lamps;

- sodium vapour lamps, of high pressure and high pressure mercury lamps with fluorescent bulb, can be used together for general lighting, only if the two types of lighting sources are placed in a lamp, special built for this purpose.

When *choosing the lamps*, must be taken into account:

- lamp photometric characteristic, so as to obtain a good total or direct efficiency of the lighting installation on the work general plan and environmental quality light conditions, stipulated by the occupational standards;

- the protection degree ensured by the lamp construction, to does not determine because of working environment, a rapid wear or to lead at undesirable events;

- maintenance requirements;

- lamp aspect, for its integration in the room space.

Regarding *the photometric characteristic choice* of lamp, it will take into account the issues that will be discussed below. In terms of light distribution compared to the horizontal plane that sectioning the lamp through its luminous centre, conventionally was established the following classification:

- lamps with direct distribution, where the 90-100% of luminous flux is distributed in the lower hemisphere and the rest is distributed in the upper hemisphere;

- lamps with semi-direct lighting distribution in which 60-90% of the luminous flux is distributed in the lower hemisphere;

- lamps with direct-indirect distribution, in which 40-60% of the luminous flux is distributed in the lower hemisphere;

- semi-indirect distribution lamps in which 10-40% of the luminous flux is distributed in the lower hemisphere;

- lamps with indirect distribution, where the 0-10% of the luminous flux is distributed in the lower hemisphere;

For achievement of the general uniform or general localized lighting, of the industrial interior spaces, are used only lamps, with luminous flux direct or semi-direct distribution. This permits the obtaining on the general plan of work, a total efficiency of the lighting installation of at least **0.6** and recommended values of the ratios between illuminations ceiling, walls and working general plan. The total efficiency of the lighting installation on the general plan is calculated with 18 relation:

$$Q = \frac{E_t x S}{F} \; ,$$
(18)

where:

Q – efficiency of the lighting installation on the working general plan;

E_t – total average illumination (direct and reflected) [lx], 1 lux = lumens per square meter;

S – surface of the working general plan [m^2];

F – luminous flux given by the installation lamps.

For the outdoor workspaces lighting, is used only lamps with direct distribution of luminous flux, to obtain on the general plan, a direct efficiency of at least **0.8**. The direct efficiency of lighting installation on the general plan is calculated with 19 relation:

$$Q = \frac{E_d \times S}{F} \ ,$$ (19)

where:

    Q – efficiency of the lighting installation on the exterior general plan;

    E_d – direct lighting on general plan [lx];

    S – general plan surface [m^2];

    F – luminous flux given by the installation lamps.

The local lighting is realized only by the lamps with direct distribution of luminous flux, specially constructed for this purpose, equipped with shielding light sources devices, so that, they do not constitute sources of disturbance of the visual field. The direct distribution lamps ensure the highest values of the system efficiency on the general plan, but can produce accented shadows, especially those with concentrated distribution of luminous flux.

Improving the lighting quality ambience, can be obtained by ensuring a superior component of the luminous flux of up to 10% through openings at the top of the reflector. In use, these allow, an air flow from bottom to top, thus reducing dust deposits on the lamp and reflector.

The direct illumined lamps with wide distribution in the lower hemisphere, equipped with optical devices (reflectors), reduce lamp luminance in the direction of view, provide a good illumination on vertical surfaces and can be mounted at a distance from each other by 2-4 times greater than their mounting height above the general plan.

The direct illumined lamps with medium to large distribution are indicated for interiors lighting with heights up to 10 m, where there is the vertically work surfaces, or luminous environmental quality requirements are high. The shadows are weak, it provides a good outlining of objects and visual field structure is comfortable.

The direct illuminated lamps with medium distribution of the luminous flux in the lower hemisphere are suitable for the halls lighting up to 10 m high, when majority work surfaces are in the horizontal plan, and the luminous environment quality requirements are not raised. It is provided a good outlining of the objects and details, of visual field.

The lamps of direct illuminated with concentrated distribution or concentrated towards average, in inferior hemisphere are indicated for spaces lighting with heights greater than 10 m, where the quality requirements for the luminous environment are low, or where there is not require, of luminous environment quality. The vertical surfaces are poorly illuminated.

The direct illuminated lamps with highly concentrated distribution of luminous flux (projectors) are useful for the exterior spaces lighting on large surfaces. Produce accented shadows, which requires attention to their placing, especially in areas with materials stored in stacks.

The direct illuminated lamps with semi-direct distribution, which provides between 10-40% of luminous flux in the upper hemisphere (ceiling, walls, floors, equipment), painted in light colours are used for height rooms up to 10 m.

The accentuated ceiling illumination reduces the contrast between the lamp luminance and ceiling luminance, ensuring a bright environment, visual comfortable. When the lamp luminance is high, the luminous flux distribution in the upper hemisphere should be increased.

Choice of the lighting mode is based on the following systems:
- uniform general lighting;
- localized general lighting;
- uniform general lighting supplemented with local lighting.

The systems of general illuminated consist of sources located to the ceiling level, of the beams or it descended by these to up to 1/3 of the height of the room. For *local lighting,* the sources will be located near to the work area.

The uniform general lighting ensures a constant average illumination, on the entire working area of room; the lamps are placed regularly in the lighting installation plan. The distance between the lamps, and the distance between the rows of lamps it maintains constant. This placement provides an ordered structure and aesthetics of the visual field. The lighting type is suitable for spaces where lighting requirements are identical for all jobs, or where machines or tables position are changed once with product manufactured, in places which require special hygiene (food and pharmaceutical industry) as in spaces congested with work equipment.

Localized general lighting is imposed in the situations where must be realized the visual tasks with a certain difficulty, in a well-defined area of the room. The main characteristic of this lighting system is that the average lighting level is changed from one area to another, on the work general plan. Localized lighting is realized either by changing the lamps density of bodies in different parts of the lighting plan, either through changes the number of bulbs lamps in the lighting fixture, their power or even the type of lamp. For the works with great visual difficulty, when is imposed as the very fine details to make highlighting, through focusing, directing or light reflection and by increasing at very high values of illumination, general illumination is completed with local illuminated.

Local lighting is realized by incandescent or fluorescent lamps, which are placed on the work surface or close to it. For temporary work, the lamp can be mounted on a mobile support. If the work requires the extremely fine detail tracking, lighting installation can be completed with optics magnifying devices (magnifying glass, microscope).

In *the natural lighting case*, the light source is solar radiation that enters in the workspaces through the gaps (windows, skylights) practiced in the building elements. Natural light is characterized by permanent change, depending on the sun position, season and cloud coverage degree of the sky. The natural lighting design is not considered a certain value of illumination, but the ratio between of interior and exterior illumination, known as the daylight factor or factor of natural lighting.

The daylight factor depends by the glazing solution established for a given building. Their choice should focus on optimization of the building global energy consumption, considering consumption for artificial lighting, heating, air conditioning and ventilation.

If the glazing optimal solution does not ensure the necessary illumination level for the unfolding of the activities, natural lighting is supplemented by permanent or temporary artificial lighting during the day.

The types of goals used to natural light, are the windows and skylights. Natural lighting through the windows is characterized by a rapid illumination decrease on the work general plan with increasing distance from them. An acceptable uniformity of illumination on the room general plan is obtained only when the distance between the windows and the opposite wall is equal to the maximum height from the floor to the top of the window. The variation of the illumination depends on the windows size, the height at which is situated their top part, the distance between them, the orientation given by the cardinal points and the lighting by reflection from the interior surface and from the neighbouring constructions.

The conclusion is that for the same area, a narrow window and high is better than one wide and low. In other words, when the area of sky seen through the window is larger, and the natural lighting from the outside is higher.

Skylight replaces a part of a roof, consisting of transparent or translucent panels (glass or plastic), which allow natural light to enter in rooms. From the point of view of the vertical section shape, are different types of skylights:

- plane – in which windows are placed in the same plane as the roof cover;

- triangular – constituted of windows arranged in two planes that forming an angle pointing up;

- rectangle – when windows are placed in two parallel vertical planes;

- shed – windows are disposed on a single side, the roof skylight can be flat or curved;

- cupola – with glass reinforced polyester.

Depending on the requirements of the technological process, the skylights are placed parallel to the longitudinal axis of the building or transversely on it. An acceptable uniformity of the illumination on the work plane is obtained if the distance between two neighbouring skylight is less than 2 x H (H – height of the skylight placement above the working plane). For greater distances between skylights is necessary to maintain in function, a part of the general lighting installation or local lighting installations.

A problem of the natural lighting is the direct sunning that, besides raising the inside temperature, can cause glare and eyestrain. To avoid these drawbacks are provided curtains, drapes, roller, blinds, insulating glass, coloured or translucent or different shading elements that belong to the construction. The skylight presence, do not exclude using of windows, because these ensure direct visual contact with the outside, which is important from the psychological aspect.

Mixed lighting – where natural light is not sufficient for visual requirements of the activities that take place in a space, it is maintained the partially artificial lighting installation. For mixed lighting no are restrictions on the combined using of natural and artificial light sources. Natural lighting can be completed by partially using of the general lighting or local lighting. Typically, the lamp rows are placed parallel with windows or skylights rows. Artificial lighting will be deactivated when the illumination on working plane is at least twice the prescribed lighting.

Safety lighting is necessary to avoid remaining in dark of the workspaces, at the accidental interruption of power supply of the normal lighting system. The emergency lighting supply is made from the central storage batteries, from an engine group – generator group or individual batteries of each lamp. Depending on the scope, exist:

- emergency lighting for the personnel evacuation;
- emergency lighting for interventions;
- emergency lighting for work continuation.

In the case of evacuation lighting, the functioning entrance time of this installation should be up to two seconds. The power supply should be designed to ensure a lifetime of at least one hour. Number of lamps is established for an illumination level, in escape ways axis, of:

- 0.5 lux in rooms with danger of explosion or poisoning;
- 0.3 lux in other rooms;
- 0.3 lux outdoors spaces with explosion or intoxication hazards;
- 0.2 lux in other outdoor areas.

For intervention lighting, operating time is still one hour, and the lighting level in areas where are action devices, must be at least 5 lux. In the case of the lighting, of work continuation, this it is sized and positioned so to ensure on the working surfaces, of 10% of the normal illumination. The duration of the entry into service is up to 15 seconds, and the lifetime of at least 3 hours. Automaton switching devices make the transfer from normal supply with electrical energy to the supply for safety lighting, when, safety lighting does not work in same time with the normal working lighting.

The safety lighting of the interior and exterior workspaces is usually done with incandescent electric lamps. Fluorescent lamps can be used for interior spaces that do not require immediate and surely entry into service of installation all lamps. Safety lighting is realized with direct type lamps with a medium or medium to large distribution of the luminous flux in the lower hemisphere. The practiced lighting systems are:

- localized general lighting along evacuation routes for the evacuation safety lighting installations;

- localized general lighting or local lighting, for the safety lighting installation, of intervention;

- uniform general lighting and localized general illumination for the safety lighting in the case of work continuation.

The lamps for safety lighting, in evacuation case, are mounted usually at a height of 2.5 m from the surface to be illuminated, along the evacuation way and above the doors, so as evacuation way stand out clearly in building conformation when the working normal lighting is shut down. Recommended height must be respected because in the fire case, the smoke and fog accumulated in large quantities to in the upper levels of the room will decrease the efficiency of the safety lighting system when increasing the mounting height.

Lighting design criteria – for a good luminous comfort, is essential as in completing the requirements of lighting, the need of quantity and quality, to be ensured. The workplace luminous comfort is ensured, if the three basic requirements of human beings are respected:

- visual comfort, which will determine the well state of the operator and indirectly, contributes to labour productivity growth;

- visual performance that will determine the visual tasks effectuation of operators, even in difficult circumstances and long intervals;

- safety.

The main parameters, determinants, of the environment illumination are: illumination distribution, lighting, glare, light direction, gleam and daylight. The all surfaces illuminations are important and will be determined by the reflection and lighting on them. For the majority interior surface, regular reflection ranges are:

- ceiling (cover) 0.6 ... 0.9;
- walls 0.3 ... 0.8;
- work plans 0.2 ... 0.6;
- floor (platform) 0.1 ... 0.5.

Staircase lighting has the following characteristics:

- increasing factor for this staircase is to about 1.5;
- illumination values in lux are: 20 – 30 – 50 – 75 – 100 – 150 – 200 – 300 – 500 – 750 – 1000 – 1500 – 2000 – 3000 – 5000.

The illumination value will be adjusted by at least one step on this scale, if visual conditions differ from those assumed to be normal. Lighting should be increased when:

- vision field is critical;
- correction of errors is expensive;
- accuracy and high productivity are of paramount importance;
- worker visual ability is below normal;
- visual task details are small or have a low contrast.

Lighting should be reduced when:

- visual task details are usually large or the contrast is high;
- visual task is for short time.

Table 4

Illumination conjunction of working area with of adjacent area

Task illuminance [lx]	Illuminance of immediate surrounding areas [lx]
≥ 750	500
500	300
300	200
≤ 200	$E_{\text{working task}}$
Uniformity: ≥ 0,7	Uniformity: ≥ 0,5

For the work areas with the continue activity, their illumination can not be less than 200 lux. Illumination of the work area will be made as uniform as possible and the illumination of the adjacent areas, to be in conjunction with its illumination (see table 4).

The glare is the sensation produced by the brightness areas there are in visual field and can be a source of visual discomfort or visual disability. It is important to limit glare (brightness) to avoid errors, fatigue and accidents. Assessment of discomfort caused by the glare (brightness), generated by the light sources that form the illumination interior system, can be determined by calculating the value of Unified Glare Rating (UGR), with the 20 relation:

$$UGR = 8 \cdot \log_{10}\left(\frac{0,25}{L_b} \cdot \sum \frac{L^2 \cdot \omega}{p^2} \right) , \qquad (20)$$

in which:

L_b – background luminance in [cd x m^2], calculated as E_{ind} x π^{-1}, where E_{ind} is the indirect vertical illumination that reaches the observer's eye;

L – illumination produced by the luminous part of each light source in the direction of the observer's eye [cd x m^2];

ω – solid angle of the luminous part of each light source;

p – Guth position index for each light source.

Brightness light sources can cause blindness and can affect the objects view. The effect is reduced or avoided by installing shields (protection screens) for light sources, or curtains for windows.

Lighting parameters measurement is made at putting into operation objectives and when performing upgrading lighting, of existing units. In such cases, it is necessary to determine parameter values that characterize the luminous ambience for the working normal and safety lighting (see Appendage

2). Illumination, surfaces luminance and their reflection coefficients are measured with apparatus, special designed for this purpose. The other parameters are determined on basis of the values obtained from measurements:

- coefficients of the illumination uniformity on the general plan and the work surface;
- reports between the main surfaces illuminations, of room;
- luminance reports.

Values obtained by measuring are valid only to existing conditions while performing its. Therefore, it is important to make a detailed description of factors that could affect the results: the light source type and the lamp type, network voltage in measurement time, reflectance of interior surfaces, maintenance program, last cleaning date and the measure apparatus with that measurement is made.

In the Appendage 2 is shown the light necessary for various activities and workspaces ($\overline{E}_m$ – light necessary, UGR_l – boundary Unified Glare Rating, R_a – colour rendering indexes).

The colour appearance of the lamp refers to the apparent colour (chromaticity) of the light emitted. It is quantified by its correlated temperature (T_{CP}). Colour appearance may also be described as in table 5.

Table 5

Lamp colour appearance groups

Colour appearance	Correlated colour temperature T_{CP} [K]
Warm	Below 3,300
Intermediate	3,300 to 5,300
Cool	Above 5,300

Offices can accommodate one or more work station in known or unknown arrangements [16]. A work station area includes desktop surface (s) and user space. The working plane is assumed to be 0.75 m above floor level. In the case of office with single workstation is known. The surrounding area is taken to be the rest of the room less a 0.5 m wide marginal strip (see fig. 11). In the case of office with unknown arrangement of work stations, the work station area should be taken as the whole room less 0.5 m wide marginal strips, which is ignored. Where planning documents shows work stations close to windows, a correspondingly wide strip can be taken as the work station area. The rest of the rooms less the ignored 0.5 m marginal strip is considered to be the surrounding area.

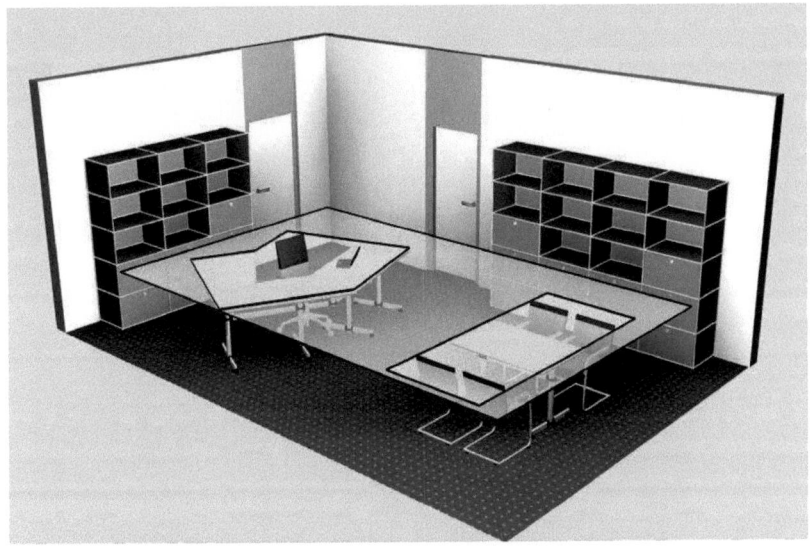

Fig. 11 *Office work station area: display screen work (medium yellow), meeting table (medium gray) and surrounding area (dark gray)*

In very large rooms where work stations are occasionally or regularly not manned (in a call centre), standard allows a background area to be applied (see fig. 12). It should be seen as a strip at least 3.0 m wide. The maintained illuminance required for surrounding and – where applicable – background areas depends on the requirements that need to be met in the work station areas.

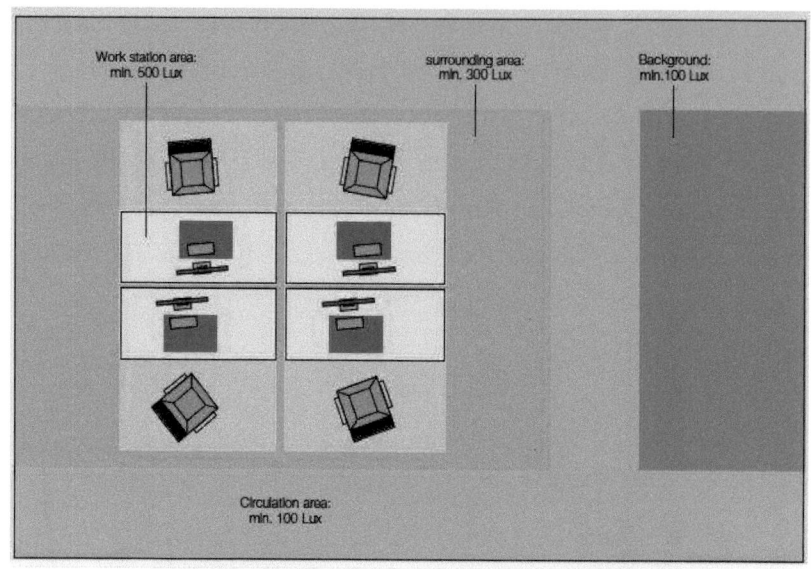

Fig. 12 *Illumination level in very large rooms*

Fig. 13 *Horizontal and vertical lighting in a classroom*

Vertical illuminance in the main viewing direction should be E_V > 100 lx in classrooms with 300 lx illuminance and E_V > 175 lx in evening classrooms and lecturer theatres with 500 lx illuminance [16]. These requirements for compliance with ASR A3.4 also apply to walls with charts and posters. No requirements are specified for individual student desks (see fig. 13). 500 lx vertical illuminance needs to be maintained over the whole surface of a chalkboard. A strip extending to each side of the board at a writing height of 1.2 – 1.8 m is used as a reference for 0.70 uniformity. Uniformity over the entire work surface should be 0.60 m.

In principle, the grid required to determine average illuminance and uniformity depends on the size and shape of the reference surface considered. Reference surfaces are work station, surrounding and background areas, on the one hand, and activity or interior areas, on the other.

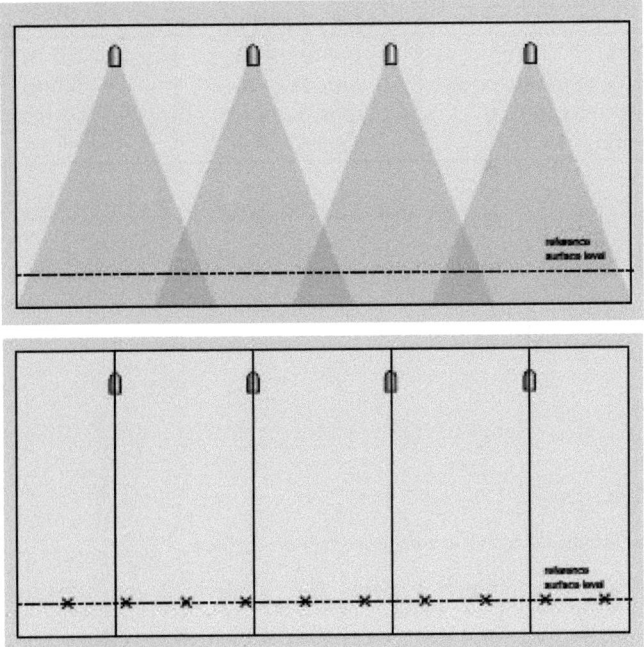

Fig. 14 *Lamps and measurement points arrangement*

Consideration needs to be given here to the geometry of the lighting installation, the luminous intensity distribution of the luminaires, the degree of precision required and the photometric quantities to be evaluated. The arrangement of luminaires and the arrangement of measurement points should not be identical (see fig. 14). The spacing between measurement points needs to be less than the mounting height. In high bays, light beams should overlap at height and not just on the reference surface. A 0.5 m wide strip along the walls is excluded from the calculation area. This is unless task areas are located within the strip or extend into it. For rooms and areas the recommended grid is presented in table 6.

Table 6

Size of grid recommended for rooms and area

Room type	Longest dimension of are or room	Grid size
Task area	approx. 1 m	0.2 m
Small rooms/room zones	approx. 5 m	0.6 m
Medium size rooms	approx. 10 m	1 m
Large rooms	approx. 50 m	3 m

Experience has shown that the following grid size p should not be exceeded [16]:

$$p = 0.2 \cdot 5^{\log_{10} d} \quad , \tag{21}$$

where:

p – grid size;

d – relevant dimension of the reference surface.

Rectangular reference surfaces are subdivided into smaller, roughly square rectangles with the calculation points at their centre. The arithmetic mean at all the calculation points is the average illuminance. Where the reference surface has a length-to-width ratio between 0.5 and 2.0, the grid size p and there fore the number of points can be determined on the basis of the longer dimension d of the reference area. In all other cases, the shorter dimension needs to be taken as the basis for establishing the spacing between grid points.

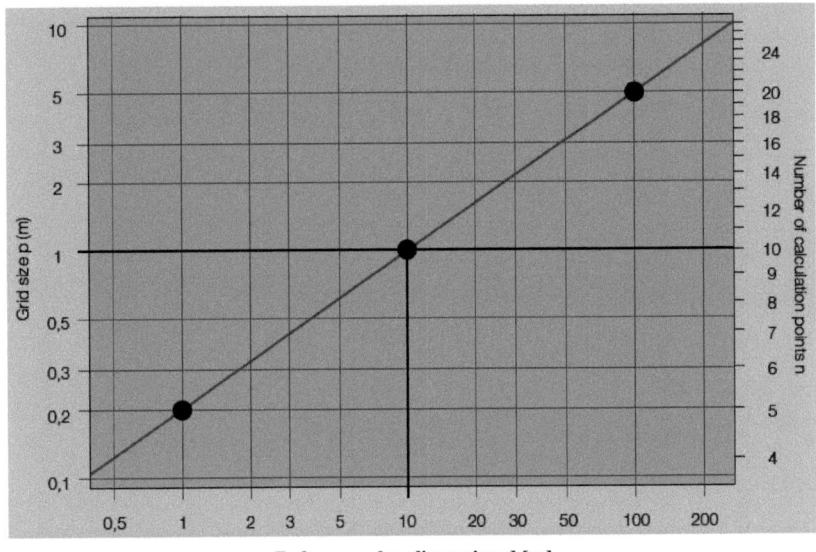

Reference plan dimension d [m]

Fig. 15 *Number determination, of calculation points*

For non-rectangular reference surfaces, i. e. surfaces restricted by irregular polygons, grid size can be determined analogously using on appropriately dimensioned circumscribing rectangle. Arithmetic means and uniformities are then established taking only the calculation points within the restricting polygons of the reference surface.

The maintenance factor (MF) is a multiple of factors and is determined as fallows:

$$MF = LLMF \cdot LSF \cdot LMF \cdot RMF \quad , \tag{22}$$

where:

LLMF – lamp lumen maintenance factor;

LSF – lamp survival factor;

LMF – luminaire maintenance factor;

RMF – room maintenance factor.

In many cases, a lamp survival factor *LSF=1* can be assumed because the failure of individual lamp leads to unacceptable falls in lighting level, which is why individual lamp replacement is required. Individual maintenance factor values can be obtained from manufactures or found in manufacturer independent standard average value curves.

Maintenance factors and conditions – where one or more of the following – potentially inter-impacting – conditions applies, maintenance factors can generally be increased:

MF – 0.80

- use of lamps subject to little light depreciation (depending on burning life);

- use of luminaires with little tendency to collect dust;

- use of operating gear that lengthens lamp life;

- short periods of service per year;

- low switching frequency;

- short cleaning and/or maintenance intervals, individual and group lamp replacement;

- low exposure to dust in the atmosphere;

- low tendency to collect dust and/or for reflecting surfaces to become discoloured;

MF – 0.67

- use of lamps subject to marked light depreciation;

- use of luminaires with tendency to collect dust;

- long periods of service per year;

- high switching frequency;

- long cleaning and/or maintenance intervals, only group lamp replacement;

- high exposure to dust in the atmosphere;

- tendency to collect dust and/or for reflecting surfaces to become discoloured.

MF – 0.50

Where one or more of the above – potentially inter-impacting – conditions applies, maintenance factors generally to be lowered.

For rough projection or where detailed information is not available, one of values from table 7 can initially be selected.

Table 7

Maintenance factors values

Maintenance factor	New-value factor	Example
0.80	1.25	Very clean room, low use installation
0.67	1.50	Clean room, 3 year maintenance cycle
0.57	1.75	Interior and exterior lighting, normal environmental pollution load, 3 year maintenance cycle
0.50	2.00	Interior and exterior lighting, dirty environment

Use of the above values does not release designers from their documentation obligation. A maintenance factor of 0.67 is recommended for comparing lighting designs maintenance.

Accessories for noise combating to receptor

Intern antiphons (corks)

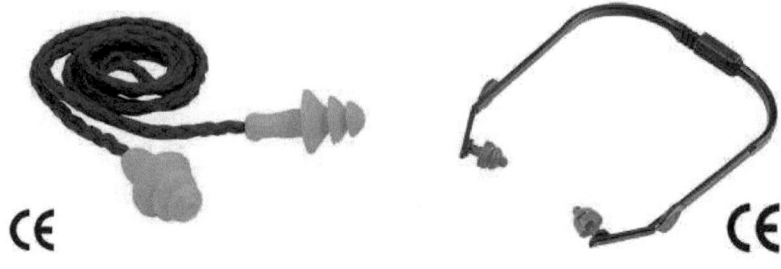

Extern antiphons

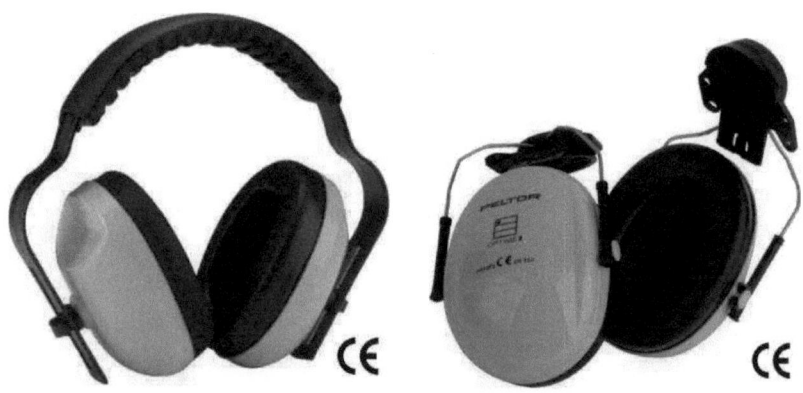

Lighting requirements for interior (areas), tasks and activities (partial)

Type of interior, task or activity	$\overline{E}_m$ [lx]	UGR $_l$	R_a	Remarks
Traffic zones				
Circulation areas and corridors	100	28	40	Illuminance at floor level
Stairs, escalators, travolators	150	25	40	
Loading ramps-bays	150	25	40	
Rest, sanitation and firs aid rooms				
Canteens, pantries	200	22	80	
Rest rooms	100	22	80	
Rooms for physical exercise	300	22	80	
Cloakrooms, washrooms, bathrooms, toilets	200	25	80	
Sick bay	200	25	80	
Rooms for medical attention	500	19	90	
Store rooms, cold stores				
Store and stockrooms	100	25	60	200 lx if continuously occupied
Dispatch packing handling areas	300	25	60	
Chemical, plastics and rubber industry				
Remote-operated processing installations	50	-	20	Safety colours shall be recognisable
Precision measuring rooms, laboratories	500	19	80	
Pharmaceutical production	500	22	80	
Tyre production	500	22	80	
Colour inspection	1,000	16	90	$T_{CP} \geq 4,000$ K
Cutting, finishing, inspection				
Foundries and metal casting				
Man-size under floor tunnels, cellars	50	-	20	Safety colours shall be recognisable
Platforms	100	25	40	
Sand preparation	200	25	80	
Dressing room	200	25	80	
Work places at cupola and mixer	200	25	80	
Casting bay	200	25	80	
Machine moulding	200	25	80	
Hand and core moulding	300	25	80	

die casting	300	25	80	
Model building	500	22	80	
Jewellery manufacturing				
Working with precious stones	1,500	16	90	$T_{CP} \geq 4{,}000$ K
Manufacture of jewellery	1,000	16	90	
Watch making (manual)	1,500	16	80	
Watch making (automatic)	500	19	80	
Metal working and processing				
Open die forging	200	25	60	
Drop forging	300	25	60	
Welding	300	25	60	
Precision machining: grinding – tolerance ≥ 0.1 mm	500	19	60	
Plate machining thickness ≥ 5 mm	200	25	60	
Sheet metalwork thickness < 5 mm	300	22	60	
Tool making: cutting equipment manufacture	750	19	60	
Galvanising	300	25	80	
Surface preparation and painting	750	25	80	
Rolling mills, iron and steel works				
Production plants without manual operation	50	-	20	Safety colours shall be recognisable
Production plants with occasional manual operation	150	28	40	
Production plants with continuous manual operation	200	25	80	
Furnaces	200	25	20	Safety colours shall be recognisable
Mill train, coiler, shear line	300	25	40	
Control platforms, control panels	300	22	80	
Test, measurement and inspection	500	22	80	
Offices				
Writing, typing, reading, data processing	500	19	80	
Technical drawing	750	16	80	
CAD work station	500	19	80	
Conference and meeting rooms	500	19	80	
Archives	200	25	80	
Reception desk	300	22	80	
Retail premises				
Sales area	300	22	80	
Wrapper table	500	19	80	
Restaurants and hotels				
Reception/cashier desk, porters	300	22	80	

desk				
Kitchen	500	22	80	There should be a transition zone between kitchen and restaurant
Restaurant, dining room, function room	-	-	80	The lighting should be designed to create the appropriate atmosphere
Self-service restaurant	200	22	80	
Buffet	300	22	80	
Conference rooms	500	19	80	Lighting should be controllable
Corridors	100	25	80	During night-time lower levels are acceptable
Libraries				
Bookshelves	200	19	80	
Reading area	500	19	80	
Theatres, concert halls, cinemas and trade fairs, exhibition halls				
Practice rooms, dressing rooms	300	22	80	Lighting of mirrors for make-up shall be glare-free
General lighting	300	22	80	
Educational building				
Classrooms, tutorial rooms	300	19	80	Lighting should be controllable
Classrooms for evening classes and adults education	500	19	80	Lighting should be controllable
Lecture hall	500	19	80	Lighting should be controllable
Technical drawing rooms	750	16	80	
Practical rooms and laboratories	500	19	80	
Computer practice rooms	300	19	80	
Entrance halls	200	22	80	
Circulation areas, corridors	100	25	80	
Stairs	150	25	80	
Teachers rooms	300	19	80	
Sports rooms	300	22	80	
School canteens	200	22	80	
Kitchen	500	22	80	
Tranportational areas (airports, railway station)				
Arrival and departure halls, baggage claim areas	200	22	80	
Connecting areas, escalators,	150	22	80	

travolators				
Information desks, check-in desks	500	19	80	
Customs and passport control desks	500	19	80	Vertical illuminance is important
Waiting areas	200	22	80	
Security check areas	300	19	80	
Air traffic control tower	500	16	80	Lighting should be dimmable; glare from daylight shall be avoided; avoid reflections in windows, especially at night
Testing and repair hangars	500	22	80	
Engine test areas	500	22	80	
Ticket and luggage office and counters	300	19	80	
Health care premises				
Waiting rooms	200	22	80	
Corridors: during the day	200	22	80	
Corridors: during the night	50	22	80	
Examination rooms: general lighting	500	19	90	
Examination rooms: examination and treatment	1,000	19	90	
Eye examination rooms: examination of the outer eye	1,000	-	90	
Ear examination rooms: ear examination	1.000	-	90	
Treatment rooms: dialysis	500	19	80	
Endoscopy rooms	300	19	80	
Treatment rooms: massage and radiotherapy	300	19	80	
Scanner rooms: general lighting	300	19	80	
Operating rooms: operating theatre	1,000	19	80	
Intensive care unit: examination and treatment	1,000	19	90	
Dentists: at the patient	1,000	-	90	
Dentists: operating cavity	5,000	-	90	Values higher than 5,000 lx may be required
Laboratories and pharmacies: general lighting	500	19	80	
Sterilisation and disinfection	300	22	80	

rooms				
Autopsy table and dissecting table	5,000	-	90	Values higher than 5,000 lx may be required

$\overline{E}_m$ – lighting necessary, UGR_l – boundary Unified Glare Rating, R_a – colour rendering indexes

REFERENCES

1.   **Catana D.** – *Statistical weights method to assessment global risk level*, 8th International Scientific Conference – Health, Work and Social Responsibility, University Urbaniana, Roma, Italia, 28-02.10.2010, 2010, p. 126

2.   **Catana D.** – *Risk assessment in occupational safety and health*, Publishing Lux Libris, Braşov, 2013

3.   **Christea Al.** – *Ventilarea şi condiţionarea aerului – Reglarea şi încercarea instalaţiilor de ventilare*, Vol. II, Editura Tehnică, Bucureşti, 1970

4.   **Darabont Al., ş. a.** – *Managementul securităţii şi sănătăţii în muncă*, Vol. I, Editura AGIR, Bucureşti, 2001

5.   **Darabont Al., ş. a.** – *Managementul securităţii şi sănătăţii în muncă*, Vol. II, Editura AGIR, Bucureşti, 2001

6.   **Darabont Al., ş. a.** – *Auditul securităţii şi sănătăţii în muncă*, Editura AGIR, Bucureşti, 2002

7.   **Micu D. A.** – *Contribution and researchers regarding the application on the thermo-mechanical treatments on the tools steels and their implication for work safety and health of the operator*, Ph.D. Thesis, Transilvania University of Brasov, Brasov, Romania, 2010

8.   **Voicu V.** – *Combaterea noxelor în industrie*, Editura Tehnică, Bucureşti, 2002, ISBN 973-31-2130-4

9.   *** – *Bruel & Kjær Sound & Vibration Measurement A/S – Hand-held analyzer Type 2250-L – Technical Documentation – Instruction manual*, Nærum, Danemarca, 2007

10. *** – *Bruel & Kjær Sound & Vibration Measurement A/S – Hand-held analyzer Type 2250-L – Technical Documentation – User manual.* Nærum, Danemarca, 2007

11. *** – *Bruel & Kjær Sound & Vibration Measurement A/S – Human vibration analyzer Type 4447 – Technical Documentation – User manual,* Nærum, Danemarca, 2007

12. *** – *Council Directive 89/391/CEE on the introducing of measure to encourage improvements in the safety and health of workers at work*

13. *** – *Council Directive 2003/10/EC on the minimum health and safety requirement regarding the exposure of workers to the risks arising from physical agents (noise)*

14. *** – http://cfcem.ee.tuiasi.ro/catedra

15. *** – www.ec.europa.eu/eurostat

16. *** – www.licht.de

17. *** – www.ro.wikipedia.org